WARRIOR • 137

ROYAL MARINE COMMANDO 1950–82

From Korea to the Falklands

WILL FOWLER ILLUSTRATED BY RAFFAELE RUGGERI

Series editors Marcus Cowper and Nikolai Bogdanovic

First published in 2009 by Osprey Publishing
Midland House, West Way, Botley, Oxford OX2 0PH, UK
443 Park Avenue South, New York, NY 10016, USA
E-mail: info@ospreypublishing.com

A CIP catalogue record for this book is available from the British Library.

ISBN: 978-184603-372-8
E-Book ISBN: 978-1-84603-883-9

Editorial by Ilios Publishing Ltd, Oxford UK (www.iliospublishing.com)
Page layouts and origination by PDQ Media, UK
Index by Alan Thatcher
Typeset in Sabon and Myriad Pro
Printed in China through Worldprint Ltd

09 10 11 12   12 10 9 8 7 6 5 4 3 2 1

FOR A CATALOGUE OF ALL BOOKS PUBLISHED BY OSPREY MILITARY AND AVIATION PLEASE CONTACT:

NORTH AMERICA
Osprey Direct, c/o Random House Distribution Center, 400 Hahn Road, Westminster, MD 21157
Email: uscustomerservice@ospreypublishing.com

ALL OTHER REGIONS
Osprey Direct, The Book Service Ltd, Distribution Centre, Colchester Road, Frating Green, Colchester, Essex, CO7 7DW
E-mail: customerservice@ospreypublishing.com

www.ospreypublishing.com

ARTIST'S NOTE

Readers may care to note that the original paintings from which the colour plates in this book were prepared are available for private sale. All reproduction copyright whatsoever is retained by the Publishers. All enquiries should be addressed to:

Raffaele Ruggeri, Via Indipendenza 22, Bologna, 40121, Italy

The Publishers regret that they can enter into no correspondence upon this matter.

THE WOODLAND TRUST

Osprey Publishing are supporting the Woodland Trust, the UK's leading woodland conservation charity, by funding the dedication of trees.

CONTENTS

ROYAL MARINE COMMANDO 1950–82

INTRODUCTION

On 28 October 1664 the Duke of York and Albany's Maritime Regiment – also known as the Admiral's Regiment – paraded on the City of London Artillery Ground. It was a new formation that had been raised 'to be in readiness to be distributed in His Majesty's Fleet prepared for sea service'. The parade marked the formation of the Royal Marines, a force that would go on to serve with the Royal Navy as sharpshooters, landing parties, 'discouragers of mutinies', ceremonial guards and later gun turret crews and bandsmen.

Two hundred and seventy-five years later, at the outbreak of World War II on 3 September 1939, the strength of the Corps of Royal Marines stood at about 15,000 men. In 1942, the formation of a Royal Marine Commando, which, among other raids, participated in the disastrous attack on the French port of Dieppe in August of that year, was a significant year in the history of the Corps. Within 12 months nine more Royal Marine Commandos had been raised and joined the Special Service Group. Fourth Special Service Brigade, which landed at D-Day on 6 June 1944, was an entirely Royal Marine formation, while in Burma 3 Commando Brigade fought the Japanese. Those Royal Marines not required for commando service re-mustered to the massive landing craft fleet operating in Europe and the Far East.

Demobilization began soon after the defeat of Japan in August 1945. After the war conscription for the armed forces, known as National Service in Britain, was retained, but the Royal Marines only accepted volunteers for the Corps. However the Royal Marine strength dropped from its wartime peak of 70,000 to 13,000 men, and in 1946 settled into three broad functions of Sea Service, Commandos and Amphibious operations. 3 Commando Brigade was formed, based around 40, 42 and 45 Commandos, and retained the Green Beret. At various times between 1951 and 1980 two other Commandos – 41 and 43 – were formed and disbanded.

In 1952, 3 Commando Brigade moved to Malta and was assigned to the Middle East Strategic Reserve, which saw service in the Suez Canal Zone, Cyprus (where 45 Commando developed ski techniques) and Suez (where 45 Commando carried out the first heliborne assault). In 1959, when National Service ceased, establishment strength dropped to 9,000 men, but all Royal Marines, with the exception of the band, became commando trained.

Between the 1950s and early 1980s Royal Marines would see action in Malaya, Korea, East Africa, Cyprus, Suez, Kuwait, Aden and the Yemen, north Borneo, Northern Ireland and in 1982 in the Falklands. Some of these comprised counter-terrorist or counter-insurgency campaigns, like the Malayan

Emergency, Cyprus and Northern Ireland. Some, like Aden and the Indonesian Confrontation, straddled conventional war and counter-insurgency. Korea, Suez and the Falklands were conventional campaigns with artillery and air power on call. At the time of the Falklands there were a small number of Royal Marines who had served at Suez – and, in effect, this book could tell their story.

Currently men from the Royal Engineers, Royal Artillery and Royal Logistics Corps support the operations of the Commandos; however, all men within these corps have to pass the all-arms commando course before they can become part of the brigade. This ensures that standards set by 'Royal', as the Royal Marines know themselves, remain high throughout the brigade.

A section of Royal Marines from 41 Independent Commando RM sport a mixture of British and US military clothing in Korea. The corporal (seated at centre) has the correct 'pusser's' issue battledress and his companion on the right the Dennison smock. Other men wear the button-fronted US Army sweater and M1 steel helmet. (US National Archives)

CHRONOLOGY

The Royal Marines trace their history back to 1664, but the first Commando was formed in 1942. This chronology follows the life of the Corps and the Commandos from their inception in 1942 to 2007.

1942   No. 40 Commando RM formed and deployed in Operation *Jubilee*, the raid on Dieppe in August.

1943   40 and 41 Commandos land in Sicily in Operation *Husky*. In the same year No. 41 Commando lands at Salerno, Italy in Operation *Avalanche*. The Royal Marine battalions are formed into Nos. 40 and 43 Commandos and see action in Italy, Albania and Yugoslavia.

1944	43 Commando lands at Anzio, Italy in January. On 6 June 17,500 Marines take part in the D-Day landings in Normandy in Commandos, ships and landing craft. 48 Commando lands on Walcheren in November.
1945	42 and 44 Commandos are involved in the Battle of Kangaw, Burma. Nos. 40 and 43 Commandos fight in the Battle of Lake Comacchio, Italy. Royal Marine Commandos conduct river crossings in North-West Europe.
1946	Commandos from HM Ships occupy Penang, Malaya. Nos. 42 and 44 Commandos occupy Hong Kong.
1948	Royal Marine Commandos cover the British withdrawal from Palestine and deploy in the Suez Canal Zone.
1949	45 Commando is in Egypt and Aqaba. 3 Commando Brigade moves to Hong Kong.
1950	3 Commando Brigade moves to Malaya. 41 Independent Commando is formed for operations in Korea.
1952	3 Commando Brigade moves to Malta. Presentation of first colours to the Royal Marine Commandos. 41 Independent Commando is disbanded at Bickleigh, Devon.
1953	3 Commando Brigade moves to the Suez Canal Zone.
1954	40 and 45 Commandos return to Malta. 42 Commando moves to the UK. The Amphibious School RM is moved to Poole.
1955–59	40 and 45 Commandos alternate on operations in Cyprus.
1956	5 November: 3 Commando Brigade spearheads landings at Port Said, Egypt.
1957	Elements of 42 Commando are present in Northern Ireland.
1960	HMS *Bulwark* is commissioned as the first Commando carrier. 45 Commando is moved to Aden. 42 Commando is moved to Singapore. The distinctive green beret is to be worn by all trained ranks.
1961	42 and 45 Commandos land at Kuwait. HQ 3 Commando Brigade is established in Singapore. 43 Commando is re-formed in Plymouth.
1962	The Limbang operation takes place. 40 and 42 Commandos are deployed in Brunei.
1963–66	3 Commando Brigade (less 45 Commando) participates in the Indonesian Confrontation in northern Borneo and Malaysia.
1964	41 and 45 Commandos are in East Africa. Lovat service dress is introduced.

1965–67	45 Commando on operations in the Radfan.
1967	42 Commando covers the final withdrawal from Aden. 45 Commando returns to the UK, 40 Commando takes on internal security duties in Hong Kong.
1968	43 Commando is disbanded.
1969	41 Commando becomes the first RM Commando to participate in operations in Northern Ireland.
1970	45 Commando is assigned to NATO's Northern Flank.
1971	3 Commando Brigade returns to the UK from the Far East. 41 Commando moves to Malta. 45 Commando moves to Arbroath, Scotland.
1972	The Commando Logistic Regiment is formed. Warrant rank is reintroduced.
1974	40 and 41 Commandos join UN Forces in Cyprus. The Corps' strength is at 7,000.
1976	RM detachments in RN frigates take part in the third 'Cod War' off Iceland.
1977	41 Commando (less Salerno Company) leaves Malta.
1978	The first ten-man RN frigate detachment is formed. 41 Commando carries out public duties in London.
1979	Salerno Company leaves Malta. 42 Commando deploys to Hong Kong for internal security duties.
1980	Comacchio Company is formed (renamed Fleet Protection Group RM in 2000). Elements of 42 Commando deploy to the New Hebrides (Vanuatu), 3 Raiding Squadron deploys to Hong Kong for duties against illegal immigrants.
1981	41 Commando is disbanded at Deal.
1982	3 Commando Brigade spearheads the liberation of the Falkland Islands in Operation *Corporate*. 3 Commando Brigade Air Squadron moves to RNAS Yeovilton.
1983	40 Commando deploys to Cyprus for a UN tour of duty.
1984	Detachments of 3 Commando Brigade Air Defence Troop embark ships of the Armilla Patrol. 539 Assault Squadron is formed. All ten-man frigate detachments are withdrawn.
1985–93	RM Commandos deploy on operational tours in Belize.

1987   The SBS is titled the Special Boat Service and comes under the command of Director Special Forces.

1988   3 Raiding Squadron is disbanded in Hong Kong.

1989   An IRA bomb explodes at the Royal Marine School of Music, killing 11 band ranks.

1990   Royal Marines embark HM Ships during the Gulf War.

1991   HQ Commando Forces and 3 Commando Brigade (less 42 Commando) deploys to South East Turkey for Operation *Haven*. Eastney Barracks is closed.

1992   Alliance with Barbados Defence Force.

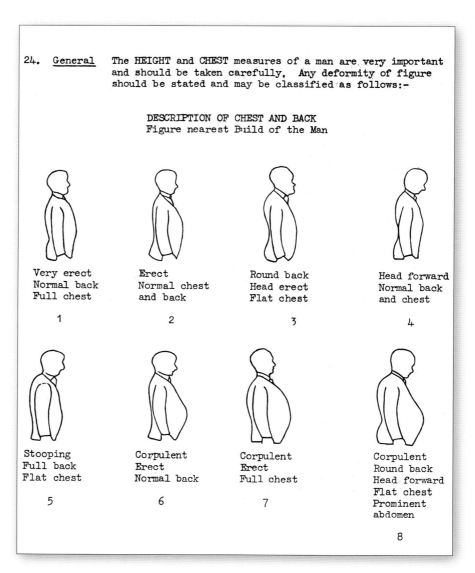

The physical types identified in *Notes on fitting of Royal Marines clothing 1955*. While it would be assumed that most Royal Marines would be those in figures 1 and 2, veterans have said that there were a few like figures 6 and 7, but none like figure 8. In fact the corpulent erect types were often pretty robust. (RM Museum)

1994	45 Commando deploys to Kuwait.

1995 HQ RM is established on Whale Island, Portsmouth. 3 Commando Brigade Air Squadron is incorporated into Naval Air Command as 847 Naval Air Squadron. The Royal Marines provide the commander and operations staff of the Rapid Reaction Force HQ in Bosnia. Commando Logistical Regiment moves to Chivenor. 42 Commando and elements of the Commando Logistic Regiment on humanitarian and disaster relief in West Indies.

1996–97 42 Commando and a detachment from 539 Squadron in the Congo prepare to evacuate civilians from Kinshasa.

1998 45 Commando is deployed on humanitarian and disaster relief in Honduras and Nicaragua. 40 Commando and a detachment from 539 Squadron in the Congo prepare to evacuate British Nationals.

1999 Ranks of Royal Marine officers are aligned with those of the Army.

2000 42 Commando deploys to Sierra Leone. HQ 3 Commando Brigade, 45 Commando, the Commando Logistical Regiment and the Royal Marine Band Plymouth deploy to Kosovo.

2001 Fleet Protection Group move from Arbroath to Faslane. New colours are presented to 40, 42 and 45 Commandos at Plymouth. Elements of 40 Commando deploy in HMS *Ocean* for operations in Afghanistan. 3 Commando Brigade Signals Squadron Royal Marine is renamed UK Landing Force Command Support Group. RM Poole is renamed 1 Assault Group.

2002 45 Commando Group deploy on operations in Afghanistan. HQ UK Amphibious Force is established with the Commandant General as COMUKAMPHIBFOR. The Corps' strength is at 6,100.

2003 The Iraq War, Operation *Telic*. 3 Commando Brigade operations in Al Faw Peninsula. Armoured Special Company RM established to develop operating procedures for Viking armoured vehicles.

2004 40 Commando conducts an operational tour in Iraq. 45 Commando completes its final operational deployment in Northern Ireland. The Viking enters service with 3 Commando Brigade.

2005 COMUKAMPHIBFOR is deployed to Iraq with his HQ to command Multi-National Division, Southeast.

2006 3 Commando Brigade is in Norway for Winter Deployment, then Afghanistan (less 40 Commando).

2007 3 Commando Brigade returns to the UK. Armoured Special Company RM (Afghanistan) is formed, and remains in theatre to support Army units. 40 Commando deploys to Afghanistan. The Corps' strength is at 6,550.

ENLISTMENT

In 1951 there were ten Royal Navy and Royal Marine recruiting offices in the United Kingdom: two in London, and one in Birmingham, Bristol, Derby, Glasgow, Liverpool, Manchester, Newcastle-on-Tyne and Southampton. Arriving at the office a volunteer would be required to present his birth certificate and attend a medical and dental examination, and to pass a simple reading, writing and arithmetic test.

It was explained to him that the Royal Marines were divided into two branches: Branch A, General Duties and Branch B, Technical. Branch A made up the majority of the Corps and included the specialities of Assault Engineer, Drill Instructor, Landing Craft, Physical Training Instructor, Swimmer Canoeist, Cliff Leader, Heavy Weapons, Parachutist, Platoon Weapons and Provost.

Branch B included the trades of Artificer (Vehicle), Armourer, Small Arms, Bricklayer, Carpenter and Joiner, Coach Trimmer (a trade that qualified the Marine as a shoemaker), Driver, Painter and Decorator, Printer, Radio Mechanic, Storeman/Clerk, technical, Vehicle Mechanic/Electrician and Metalsmith. Within Branch B were the non-tradesman group of Clerk, Cook and Signaller.

In the 1950s a man aged between 17 and 28 could enlist as a Marine for continuous service of 12 years, or enlist on a Special Service Engagement for seven years from the age of 18 or date of enlistment. The Marine could transfer to a Continuous Service Engagement in his fourth or sixth year.

This was also the period of National Service, which lasted from 1946 to 1960. Conscripts called up as National Servicemen within this period would make up 30 per cent of the strength of the Royal Marines. By the time it ended, over 300 National Service officers and 9,000 Royal Marines had served in the Corps.

The 1948 National Service Act fixed a period of 18 months full-time service followed by four years as a reservist. In 1950 the full-time period was extended to two years. The arrival of a buff envelope with the letters OHMS was the first indication to a young man that he had received his enlistment notice – he had been called up. Cynics asserted that OHMS did not stand for On His (later Her) Majesties Service but 'Oh Help Me Saviour' – the cry of the desperate young conscript.

Because National Servicemen made a decision to join the Royal Marines and with that the longer service and training period, they were often regarded as 'volunteers'. Vic Pegler who served in Support Troop 45 Commando would say of them, 'Some of the best Marines I have ever seen under effective fire were National Servicemen. The phrase "I was only a National Service Marine" has no place in my vocabulary.'

A   **ROYAL MARINES OF 41 INDEPENDENT COMMANDO LAYING DEMOLITION CHARGES DURING THE KOREAN WAR**

During the Korean War, between 1951 and 1952, men of 41 Independent Commando, such as the two figures shown here laying demolition charges on a railway line, were issued with US M1943 combat clothing and weapons. This made logistical sense, and, moreover, much of the American clothing and equipment was superior to those issued to British forces. However, the Royal Marines retained their 'boots SV' and their distinctive green berets (sometimes minus cap badges) – even during the harsh Korean winter. The sateen cotton jacket was waterproof when issued, but the treatment soon washed out. The US uniform with its four-pocket jacket, liner and detachable hood formed the basis of the 1950 combat clothing that was tested in Korea. The standing figure Royal Marine carries a US .30-calibre M1 Garand self-loading rifle, and is shown wearing US M1936-pattern webbing plus a cotton ammunition bandolier, slung across his chest.

The end of National Service coincided with the end of the battleships and cruisers in the Royal Navy. Since there was now no longer a requirement for men in 'blue beret sea service' the Royal Marines could concentrate on their Commando role.

By 1966 terms of service had changed to Long Service and Reserve Engagements. These entailed nine years in the Royal Marines if over the age of 18 on joining, or to the age of 27 if entering before reaching the age of 18. The period of active service was followed by three years' service in the Royal Fleet Reserve. While in the latter, Marines were paid a daily retainer and were liable to recall for active service in the event of a national emergency. In all other respects they led normal civilian lives.

Before completion of their first nine years' service over the age of 18, men could volunteer to re-engage for a further five years. It was during this period that they could apply for re-engagement to complete 22 years, which qualified them for a pension. For some specialists there was even scope to serve beyond 22 years.

By the early 1970s, though, the conditions of service had changed little, with a nine-year engagement now being called a Career Engagement. Also, a new method of entry had been introduced – the Notice Engagement. This carried the right to give 18 months' notice of intention to leave after 1½ years' service, counted from the age of 18 years, or on completion of training, whichever was later. It also offered the opportunity to transfer to a Career Engagement later and so to earn a higher rate of pay.

Men of 41 Independent Commando RM wait aboard a warship prior to a raid on the North Korean coast during the Korean War. The presence of two men equipped with the SCR 300 (later adopted by the British as the 31 Set) a 3.5-inch rocket launcher (Super Bazooka) and Browning Automatic Rifle (BAR) suggests that this is a troop HQ group. The troop commander is probably the man armed with an M1 (Grease Gun) SMG and equipped with binoculars. Of interest is the BAR man armed with two 36 grenades hooked rather precariously to his combat jacket pockets. (US National Archives)

TRAINING

Up until 1977, once the recruit had passed the medical and other tests, he would be sent to the depot at Deal in Kent. His world would now contract to a bed in a barrack room with a chair, and a locker with a key. Kit and valuables, it was explained, were to be kept locked securely in the locker.

At Deal, as a member of R (Recruit) Wing, he would be kitted out with uniform and equipment and taught how to look after it. The 1968 handbook for new recruits at Deal explained:

During your time at the Depot, you are issued with a basic kit of clothing, equipment and necessaries. Some of your clothing will be altered free of charge so that it fits you properly. Once your kit has been issued, its subsequent upkeep is your responsibility, and it is in your interests to care for it well. If you fail to do so, it will cost you money to replace spoiled or missing articles.

The handbook explained that a kit upkeep allowance (KUA) would be added to pay to ensure that Marines keep their uniforms in good order. It also added that kit and equipment had to be marked.

The 1973 handbook explained that at Deal the recruit would meet his troop instructor:

[He] will soon become the most important man in your life – a hard task-master, but a fair one, he is largely responsible for your discipline and training. Through him you will learn what is expected of a Royal Marine.

The recruit's education would be continued and physical fitness improved, and he would be taught to swim. In later years this phase of physical training would become more important as the upper body strength of recruits required

ABOVE LEFT During a raid against North Korean coastal rail links with China in October 1950, Royal Marines wearing the dark grey US Navy life jackets and armed with an M1 Garand and M1 SMG cover the demolitions team as they prepare their charges. The men wear US Army combat clothing, but retain their distinctive green berets. (US National Archives)

ABOVE RIGHT A shaped charge is positioned on the gravel ballast to create a camouflet cavity, which allows a larger charge to be placed, thus wrecking the track. The men of 41 Independent Commando RM are linking the detonating cord ring main, while the man in the centre holds a tin of detonators. They are armed with M1 Garand rifles. (US National Archives)

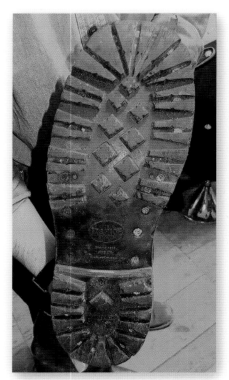

The distinctive pattern of the 'boot SV'. The cleated rubber sole was developed before the war by Vibrami, an Italian Alpinist and adopted by the Commandos during the war. The sole is screwed to a leather sole and consequently allows the leather to mould to the wearer's feet. Tough and comfortable, the boot SV was a popular item of kit. (Author's photograph)

development before they could carry weight on their shoulders over long distances in the Commando tests.

He would learn parade-ground drill and simple weapon training and be introduced to the customs, traditions and history of the Corps. At the end of this 15-week period he would be posted for a two-week course in elementary seamanship on a Royal Navy warship, for a long time HMS *Bellerophon* in Portsmouth. Aboard ship he would learn about the role of the Royal Marines at sea. By the 1970s seamanship techniques were taught at the RM Amphibious Training Unit at Poole. Here the Marine also learned about the role of the elite Special Boat Section (SBS).

After this it was down to the Infantry Training Centre Royal Marines (ITC RM), which would later become the Commando Training Centre (CTC) at Lympstone, near Exeter, Devon. He would dismount from the train at the railway halt that boasts the unique name 'Lympstone Commando'. Training in the late 1960s comprised a 17-week course consisting of an infantry course followed by the Commando course. The Commando test consisted of four tests:

1. Speed march. This six-mile (9.65km) march had to be completed in 60 minutes carrying a rifle and equipment, and finished at a rifle range with a marksmanship test.
2. Endurance test. A four-mile (6.4km) run with a rifle and equipment over bog and rough country, overcoming obstacles and crawling through tunnels, one filled with water. This test had to be completed within 80 minutes and finished at the rifle range, where marksmanship had to be proved.
3. Assault course. This was over varying obstacles, carrying a rifle and equipment, and had to be completed in five minutes.
4. Twelve-mile (19.3km) cross-country march. This was undertaken carrying 90 lb (40.8kg) of equipment.

A newly trained Marine is quoted in the 1973 handbook as follows:

It's not easy to win the green beret, but it's worth the effort when you do. The early weeks are the worst. There is no let up, exercises in all weathers. You wish at times you never began. It gets easier, though, as you get fitter.

B   ROYAL MARINE RECRUITS ON THE SIX-WEEK COMMANDO COURSE, AT THE INFANTRY TRAINING CENTRE, LYMPSTONE, DEVON, c. 1966

These four illustrations show training at the Commando Training Centre Royal Marines (CTCRM) at Lympstone. During the 1960s the majority of Royal Marine training was relocated there from Bickleigh Camp (near Plymouth). The Marines are shown climbing a 10ft wall (1), wading through water on a water obstacle course (2), during a fitness run (3), and practising the 'dead man's crawl' (4). The recruits all wear olive green denims, boots, puttees and 1958-pattern webbing. Any weapons carried during these exercises would become covered in mud and grime, but by inspection time at the close of the day they would be spotlessly clean again. In the course of their time at CTCRM the recruits would be watched closely and graded not only for their physical fitness and aptitude but also for their attitude and ability to work with the troop. The staff would identify outstanding recruits, with one in particular singled out to be 'The King's Badgeman'.

Twenty years later the Commando tests were very similar. Each man would be carrying 30 lb (13.6kg) of kit, but as the kit inevitably became soaked with water and sweat this weight increased to at least 40 lb (18kg). Each man would carry a self-loading rifle (SLR) weighing about 10 lb (4.5kg) – and thus loaded he would be up against the following challenges:

1 The Tarzan course – to be completed in five minutes by recruits and 4.5 by young officers (YOs).
2 The Tarzan course and assault course – 13 minutes for recruits, 12.5 for YOs.
3 The endurance course, plus a four-mile (6.4km) run back to camp – 73 minutes for recruits, 70 for YOs. Men would then fire on the 25-metre range.
4 A nine-mile (14.4km) speed march – 90 minutes for all. This was followed by a troop attack exercise.
5 A six-mile (9.65km) speed march – 60 minutes for all.
6 A 30-mile (48km) 'yomp' – eight hours for recruits, seven for YOs.
7 A 30ft (9m) rope climb – no time limit; however, it should have been completed in a few minutes.
8 Battle swimming test – no time limit. The man jumped into the pool wearing full kit, swam for 164ft (50m), trod water for two minutes, took off his kit and handed it to a man on the side of the pool, and trod water for another two minutes. At no time could he touch the side of the pool.

Officers and Marines trained together at Lympstone to qualify for their green beret. A veteran described it as 'a microcosm of war', since it taught and tested weapon handling and tactical skills as well as building stamina and strength.

Throughout the training recruits and YOs were under constant scrutiny, with their personal qualities being graded. These included mental ability, initiative, reliability, physical ability, professional ability, general conduct, unselfishness and leadership potential. The grades ran from 5 – very good, 4 – good, 3 – average, 2 – below average, to 1 – poor.

The training did not come cheap; in the 1980s it was estimated that it cost £30,000 to put a recruit through Lympstone and £45,000 for a Young Officer. Some men never made the grade, so there was pressure to ensure that those who did and had been tested and trained, were posted out to their Commandos.

During a break in training in the 1980s one group of YOs, obedient to a fault, followed the orders of their staff sergeant instructor and charged through underbrush chanting 'Zulu, Zulu' towards a group of unidentified uniformed men standing in a clearing. Only when they were in the clearing did they realize that they had 'ambushed' the Commandant General Royal Marines on a visit to CTC RM.

When they passed off the square, the recruits, now fully fledged Royal Marine Commandos, would be able to operate their personal weapon as well as those within the company and be able to communicate on platoon and section radios. For the bulk of the period between 1950 and 1982 these were the A41 Troop radio and C42 vehicle-mounted company and Commando radio. By the time of Operation *Corporate* – the campaign to liberate the Falklands in 1982 – British armed forces had moved to the Clansman range, which was lighter, more reliable and had rechargeable batteries.

Throughout training teamwork was (and still is) emphasized, in contrast to some military selection courses that tested the individual and emphasized only strength and stamina. Lieutenant Neil Ascherson, a National Service officer who served with 42 Commando and subsequently enjoyed a distinguished career as a writer and journalist, recalled:

> One good thing about the Marines was that we all did basic training together; there was no segregation. They knew I could do what they could do, which made them respect me.

The testing and training did not end once a Royal Marine had won his green beret. Over ten weeks he developed his theoretical knowledge and practical skill with the weapons of the infantry battalion. In the 1950s these consisted of the .303-inch No. 4 Mk 1 Rifle, a bolt-action weapon that had seen service in World War II and would equip British forces up to the 1960s. They were trained on the Mk V 9mm Sten submachine gun (SMG), a weapon that was being replaced in the late 1950s by the Sterling SMG. The anti-tank weapon was the 3.5-inch rocket launcher, which weighed only 12 lb (5.5kg) and when not in use was hinged into two sections. They learned how to clean and prime the No. 36 grenade, based on the World War I Mills bomb.

Two weapons that would span the years between 1950 and 1982 were the 2-inch ML Mortar and the .303-inch Bren light machine gun (LMG). The 2-inch Mortar Mk 7* or Mk 8 was used to fire smoke and illumination bombs and high-explosive (HE) during the Confrontation with Indonesia in the 1960s. The mortar had a maximum range of 1,640ft (500m) and when firing illuminant bombs had the advantage that light was almost instantaneous – unlike handheld rocket flares, which gave the enemy seconds to take cover as the rocket streaked upwards.

In the world's first helicopter air assault, men of 45 Commando race from their Westland Whirlwind helicopters during Operation *Musketeer* at Suez on 6 November 1956. These helicopters could hold about six Royal Marines, but aboard the tiny Sycamore helicopters three men sat with their feet hanging out of the door, with the middle man being held tightly by his two comrades. (Private collection)

The Commandant General RM (CGRM) General Sir Campbell Hardy and an aide make an unofficial visit to the 45 Commando landing zone (LZ) at Suez. The CGRM explained that he made it 'because over a third of the Corps were at Suez'. The statue of the French engineer De Lesseps, later demolished by the Egyptians, can be seen on the skyline. (Private collection)

The Bren was a box magazine-fed weapon, which, when re-barrelled for 7.62mm NATO calibre, would emerge as the L4A4 LMG. The advantage of box feed was that there was less risk of fouling or snagging that could occur with belt-fed weapons like the General Purpose Machine Gun (GPMG). This meant that the weapon had its advocates in the Corps, who added that the slower rate of fire (500 rpm) made for more accurate shooting. By now the L2A2 grenade had replaced the No. 36 grenade; the new grenade had a coil of pre-fragmented wire inside the lightweight body. When the No. 36 grenade was thrown there was a four-second delay before it exploded; it could kill or injure up to 60ft (18.29m), but on hard ground fragments could reach as far as 500ft (152.4m). In reality the No. 36 grenade often broke into large fragments, including the screw-in base plate or simply into iron dust. The L2A2 was first used in the Falklands, where the soft ground reduced its effectiveness, producing critical comments from Royal Marines.

By the 1970s the weapons in service were the self-loading 7.62mm L1A1 rifle, the 9mm L2A3 Sterling SMG, and the 7.62mm L7A2 GPMG. Though the Sterling was a robust and reliable SMG, the pistol-calibre rounds were not effective over ranges in excess of 66ft (20m). It was meant to be carried by radio operators; however, many of these opted to bear the extra weight of the SLR, since it was a more effective weapon. The SLR could fire the Energa grenade, a shaped-charge anti-tank grenade, but its short effective range made this a weapon of last resort.

The GPMG fired from 50-round, disintegrating metal belts at a rate of 1,000 rounds a minute. As a general-purpose machine gun it could be fired in the light role from a bipod or the sustained fire (SF) role mounted on a tripod. In the SF role it could reach out to 1.1 miles (1.8km).

Two new anti-tank weapons entered service in the 1970s and 1980s: the Swedish Carl Gustav 84mm anti-tank weapon and the American-designed, single-shot 66mm M7A2 rocket launcher. The weapons were known by their calibres – the 'Eighty-Four', or 'Charlie G', and the 'Sixty-Six'. The Carl Gustav, which fired a high-explosive anti-tank (HEAT) round, had an effective range against moving targets of 1,312ft (400m) and could penetrate 15.7 inches (40cm) of armour. The Sixty-Six had an effective range of 984ft (300m) and could penetrate 1ft (30.5cm) of steel. The attractive feature of the telescopic Sixty-Six was its overall weight of 5.2 lb (2.37kg) and closed length of 1ft (65.5cm), which meant that two or three could be slung over a Marine's shoulder. Because both weapons had a dangerous back blast, the drill was to warn men in the area with a yell of 'Eighty-Four!' or 'Sixty-Six!' before firing.

Marksmanship and sniping were emphasized in training. During the Troubles in Northern Ireland, Royal Marines brought their skills as snipers to the conflict. The weapon of choice was the 7.62mm rifle No. 4 Mk 1 (T), and the Mk 1* (T) – a re-barrelled No. 4 fitted with a telescopic sight. This was designated the L42A1 rifle, and using the telescopic sight a sniper could hit targets at ranges of between 2,625ft and half a mile (800m and 1,000m).

Familiarity and confidence with this range of weapons was critical for realistic training in the field. This would be built up with pairs of Royal Marines, through sections, troops, companies and finally a full Commando deployed in the field. Initially training would be conducted using blank ammunition – the SLR, GPMG and LMG being fitted with yellow blank firing attachments (BFAs) to ensure that gas was tapped back into the mechanism to make the weapon reload.

Live firing was a more challenging form of training. A section or troop would have a GPMG 'shooting them in' on an objective that would contain Figure 11 rifle targets, the outline of a running man pasted to plywood. Alternatively, live firing could have two men giving one another covering fire as they skirmished towards their objective. Eventually mortar or even artillery fire would be built into the live firing exercise to make things as close as possible to actual combat conditions. Close-quarter battle, or CQB, would see one or two Marines advancing down a range laid out in scrub or rough terrain. CQB would test reflexes and reactions, as targets would appear fleetingly from behind obstacles or terrain features and the Marine would snap off one or two shots before he dashed for cover.

Along with skill at arms, the Marine would learn unarmed or close combat – the ability to disarm and overpower a man armed with a knife or even a handgun. While it might be rare that men found themselves in this situation, it built confidence and was excellent for fitness and agility.

Living in the field, a skill that the Marine would have been introduced to early in his career, would be refined, and men would learn how to use their ponchos to erect a simple shelter for an overnight stay. The night routine of sentries and 'stand to' would be learned and rehearsed. The men would become familiar with the smelly solid fuel Hexamine cookers and one-man 24-hour ration pack with its tins and packets. These rations varied according to the theatre, with arctic and jungle packs containing different main meals and snacks. Arctic packs had dehydrated meals and were very calorie rich, since there would be plenty of snow around and thus a plentiful supply of water available for rehydration.

Officers watch as a Royal Marine from 45 Commando is lowered into a well outside Akenthou during anti-terrorist operations in Cyprus in 1956. Wells could be used by EOKA for concealing arms and equipment. (RM Museum, 7/20 8 (1))

The down time, often at the end of a day, when men could cook their rations – supplementing them with spices and condiments that an experienced Marine would have as part of his kit – were a subconscious reward for the rigours of the day. The ultimate pleasure was and remains brewing a 'wet' – a steaming mug of hot, sweet tea or instant coffee.

Along with tactics of the attack that used 'fire and manoeuvre' Marines would learn how to dig a defensive position. They would learn the importance of overhead cover to protect a two-man trench from the lethal effects of mortar and artillery fire. The principles of overlapping and interlocking fields of fire for the automatic weapons – either LMG or GPMG – would be learned.

However, all this was for war in temperate conditions like north-west Europe. The Marine would also find himself learning about the drills and skills of life in the Arctic, desert and jungle – theatres that such men would become familiar with between 1950 and 1982.

Royal Marine Murphy of 40 Commando working with a Cypriot policeman at a vehicle checkpoint (VCP) at Episkopi on 26 July 1974. Marine Murphy has a 1958-pattern belt with a 1944-pattern water bottle and non-standard ammunition pouches for his L1A1 SLR. (RM Museum, 7/20 8 (90))

By the 1960s Royal Marine Commandos knew they might be delivered to battle by day or night by helicopter, landing craft, assault boat, through arctic snows, jungle cover or by cliff assault. They might be required to mount short-duration raids, or fight as conventional troops through long campaigns. This often required a high level of personal initiative and ability to work in small groups, and as a result the Royal Marines set themselves high standards in recruit selection and training and in unit training.

The newly fledged Commando qualified for the Army parachute course without having to do the gruelling Parachute Regiment P Company course – the Commando course being considered a sufficient test of grit, strength and stamina.

Until 1999 Royal Marine officers and NCOs were often slightly older than their Army counterparts. Thus, a Royal Marine corporal equated to a sergeant in the Army, while a Royal Marine captain commanded a company, a position that is held by a major in the Army. It was believed that the benefits of slow promotion were that junior leadership was very experienced and more mature. In 1999 the ranks of RM officers were aligned to those of the Army.

DAILY LIFE

The 'daily life' of a Royal Marine in the period under study could vary enormously, and this can also be considered one of the defining features of the Royal Marines. Some men would be part of a Commando, with its infrastructure of companies and troops and officers and NCOs. In the United Kingdom this could mean permanent accommodation with barracks, married quarters, stores, garages, mess halls and parade grounds. This would be the base from which the Commandos – 40, 41, 42 and 45 – would deploy and in which they would train and prepare for operations. At various times this has been Arbroath in Scotland for 45 Commando and Stonehouse Barracks in Plymouth and Bickleigh in Devon for 40 and 41 Commandos.

Overseas Commandos were based in Singapore and Malta. Malta was the base for the Suez operation in 1956 and the Radfan withdrawal from Aden in the early 1960s. Singapore was used for the Emergency in Malaya in the 1950s and the Indonesian Confrontation in the 1960s.

Officers and NCOs would have their separate messes and the Marines their cookhouse. The Navy, Army and Air Force Institute, universally known as the NAAFI, was a combination of shop, pub and club.

At these permanent bases married men would have quarters for their wives and families. Those for Marines resembled a well-run housing estate, in which the quality of the housing improved with the rank of the occupant. The commanding officer would have a house large enough to house visitors and guests and offer the formal and informal hospitality that knits a military formation together. When the Commando deployed on exercise or operations the families would remain in these bases, giving one another mutual support.

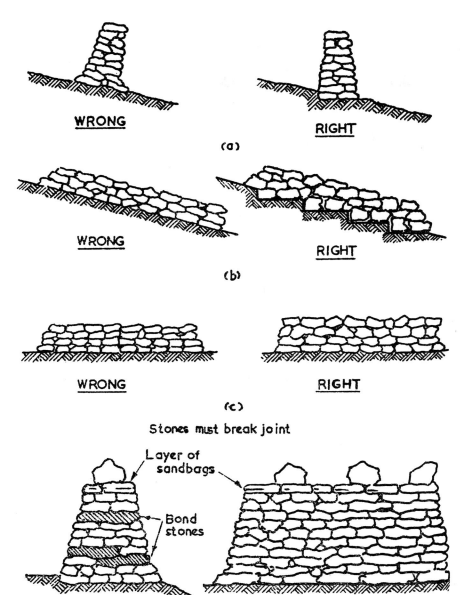

WRONG **RIGHT**

(a)

WRONG **RIGHT**

(b)

WRONG **RIGHT**

(c)

Stones must break joint

Layer of sandbags

Bond stones

SECTION **ELEVATION**

(d)

The correct and incorrect ways to build a *sangar*. In an ideal situation sandbags would be used to anchor the rocks, and racks placed at random intervals along the parapet to break the line and conceal anyone looking over the top. Sandbags are positioned on the parapet to reduce the risk of fragments from rocks chipped by small-arms fire. (Author's collection)

For Marines an ordinary working day consisted of ten hours, with the evening free unless training required men to work at night. Breakfast was normally between 07.30 and 08.30 hours, a midday meal between 12.00 and 12.45 with tea between 15.30 and 16.15, though a combined tea/supper meal could be served between 17.00 and 18.00. Officers would normally have a light tea meal with dinner being served between 19.00 and 20.00.

All ranks qualified for leave, which varied in length from a weekend to a long leave of 14 days granted at Christmas, Easter and mid-summer, and sometimes longer periods depending on whether a man was between postings. In the 1950s Marines in training had strict instructions on how to conduct themselves when on leave:

a) Do not do anything that might bring disgrace to your Corps.
b) Always be properly dressed.
c) If with a serviceman, walk in step.
d) Pay the proper compliments to all officers, whether they are in uniform or plain clothes.
e) Never walk more than two abreast.
f) Behave in an orderly manner. Do not shout or sing.
g) If on long leave, READ THE INSTRUCTIONS on the back of your leave pass.

Uniform could be worn on leave, until the Troubles in Northern Ireland made any servicemen wearing uniform in public potential targets. Initially battledress and later the distinctive Royal Marine Lovat green service dress. When Royal Marines returned to Stonehouse Barracks in Plymouth after the Falklands campaign in 1982, local girls were happy to be seen out with a Marine in his distinctive Lovat green service dress, when the restrictions were briefly relaxed.

The Royal Marines have a rich and diverse argot (see the *Glossary*) but in the 1950s and 1960s the local population of Plymouth had one for the newly promoted lance-corporal. Delighted by this preferment the young man would celebrate in town on a 'run ashore' and then take a taxi back to Stonehouse Barracks. If, worse for wear, he encountered the orderly sergeant at the gates, his career as a lance-corporal would be short lived. He would be 'busted' back to Marine the following morning by his company commander. For the locals he was therefore 'a taxi corporal'.

A Westland Wessex helicopter brings in a 105mm pack howitzer as an underslung load to a mountain position in the Radfan. The Italian-designed pack howitzer would give invaluable service both in the Radfan and the Indonesian Confrontation. Designed for mountain operations, it could be broken down into loads that could be carried by mules. (Author's collection)

An A41 radio operator with X Company, 45 Commando conducts a radio check with fellow operators before the commando moves out into the Radfan. The radio used at troop level was rugged and reliable, but was heavy and required disposable batteries, giving it total weight of around 50 lb (22.6kg). (RM Museum)

Canoeists of the Special Boat Service (SBS) in a training exercise during operations in Borneo. The ability of SBS patrols to penetrate narrow coastal waterways enabled the British and Malaysian forces to keep a close watch on Indonesian movements and to dominate the border area. (RM Museum)

Other Marines might find themselves in small troop-sized or smaller naval parties aboard ships; perhaps the most famous is NP8901, which formed the garrison on the Falkland Islands.

In 1960 the aircraft carrier HMS *Bulwark* (known to sailors and Marines as 'the rusty B') took on new life as a Commando carrier. In an innovative move she embarked 42 Commando with attached commando-trained gunners and engineers. While they could make a conventional amphibious landing on a beachhead, the versatility of the Commando carrier was that the helicopters on board could be used to insert a battalion-sized force behind a hostile shore in just two hours. HMS *Bulwark* was joined by HMS *Albion* in 1962 and the two Commando carriers were stationed either in the Mediterranean or the Far East. *Albion* was taken out of service in 1973 and replaced by HMS *Hermes*, which served for three years. 'Rusty B' was the longest-serving Commando carrier, finally being withdrawn from service in 1984. During their service the carriers provided a self-contained 'home' for the Royal Marines that could be moved to potential trouble spots if required. They would be the basis of HMS *Fearless* and *Intrepid*, both landing platform dock (LPD) vessels.

With the prospect of an operational deployment a Commando would 'work up', concentrating on refining the drills and skills that would save lives and allow them to perform effectively. Northern Ireland (or 'NI') training at the ranges at Hythe and Lydd in Kent was a regular feature of the routine of all Commandos during the 1970s and 1980s.

In addition to NI training during these years challenging, more congenial work-up would take place in preparation for winter warfare training in Norway. Over alternate years Commandos would participate in two exercises in Norway on NATO's Northern Flank: *Mainspring* in the spring and *Clockwork* in the winter. In addition to these exercises Commandos would have their own specialities; thus, 45 Commando concentrated on mountain and arctic warfare and 40 Commando on jungle training.

For married men the pressure of operational tours and training made difficult demands on their relationships. In the 1980s it was estimated that in

In the damp heat of the jungle the canvas and rubber jungle boot had a nasty habit of rotting at the join between the uppers and the sole. It was, however, in many ways a well-designed bit of kit that gave support and protection to the lower leg and foot against leeches and other jungle parasites. (Author's collection)

OPPOSITE PAGE A Fleet Air Arm Westland Wessex helicopter winches a soldier down into a clearing in Borneo during the Indonesian Confrontation. The Royal Marines had developed an excellent working relationship with the helicopter pilots and pioneered many of the techniques that became standard operating procedures (SOPs). In the foreground an officer armed with a 9mm Sterling SMG watches the perimeter of the LZ. (MOD)

a year between tours in Northern Ireland and training for Norway the men of 45 Commando spent fewer than six weeks at home. Married or accompanied postings were allowed only if the posting exceeded 12 months.

At their Commandos, specialist training for men in the General Duties branch would begin. This could be as an assault engineer; like the assault pioneers within a British Army infantry battalion, these men would include in the skills the ability to locate, lift and neutralize mines – and lay them – and the construction of field fortifications and obstacles and demolitions. In training they would fire the explosive charges for battlefield simulation, commonly called 'bat sims'. The assault engineers would learn the skills of using 'snowcrete'– snow, frozen hard and reinforced to produce material that could withstand small-arms fire and shell fragments.

Other important specialist training included men attending the heavy weapons course. In the 1950s these men would have learnt how to handle the 3-inch mortar, the .303-inch Vickers medium machine gun (MMG), and the 120mm L4 mobat and L6 wombat recoilless anti-tank guns. By the 1980s the weapons had changed to the 81mm L16 mortar, the GPMG in the sustained fire role and the MILAN wire-guided anti-tank missile. The Vickers MMG could trace its origins back to World War I. A water-cooled machine gun, it was heavy and had a cyclic rate of 500 rpm. However, it was robust and very reliable and could hit area targets at 3,800 yards (3,475m). Like the support company within an Army battalion, the heavy weapons provided organic fire support greater than that available within the troop.

If heavier weights of fire were required, the guns of 29 Commando Regiment, Royal Artillery would be called up. In the 1960s, 29 Commando Regiment – a Royal Artillery regiment composed of Army gunners who had passed the All Arms Commando Course – were equipped with the 105mm pack howitzer; by the time of the Falklands campaign, they had the versatile 105mm light gun. Likewise the assault engineers could call on assistance from 59 (Independent) Squadron Royal Engineers, Commando-trained sappers who would bring their specialist kit and experience to an engineer task.

If the Royal Marine had qualified for promotion to corporal or higher he would be eligible for training as a drill instructor, helicopter pilot, Royal Marine provost or physical training instructor. He would be posted away from the Commando to the appropriate Army, Royal Navy or Royal Marine training establishments.

Though many of the skills and trades that Royal Marines developed in their time within the Corps or their respective Commandos were similar to or even identical to those of the Army, there were two that were unique and would prove invaluable in 1982: landing craft crew, and swimmer canoeist. Though the latter sounds innocuous, it was the qualification for the SBS –

initially the Special Boat Section and later the Special Boat Squadron. Among the roles assigned to the special forces of the SBS were beach reconnaissance – the work undertaken by the Combined Operations Pilotage Parties (COPP) in World War II.

The skills developed by the Mountain and Arctic Warfare Cadre from the Commando Cliff Assault Wing RM put them very close to also being special forces. There is a challenging selection process and tough training regime to qualify men as 'mountain leaders'. In the Falklands they became the reconnaissance and intelligence gathering asset for 3 Commando Brigade HQ, and performed an invaluable role in the campaign.

The landing craft crew would train at Poole, the large natural harbour on the Dorset coast in southern England. Their amphibious warfare skills proved critical to the success of the Falklands campaign.

One of the factors that helped British forces to victory in the Indonesian Confrontation was winning the trust and confidence of the population living on the porous borders between Sabah, Sarawak and Indonesian Borneo. Known as 'hearts and minds', it included bringing health checks and assistance to the population. Here a Royal Marine medic conducts a dental check. (RM Museum)

APPEARANCE AND DRESS

In the field the appearance and dress of Royal Marine Commandos has been, and still is, dictated by the weapons issued and the theatre in which men are operating. There is one universal: wherever practicable, men wear their green Commando berets with the globe and laurel cap badge – even during the bitter winter months of the Korean War.

In the 1950s, during the campaigns in Korea, Cyprus and Suez, Royal Marines were issued with equipment and weapons that in many cases dated back to World War II. As the name suggests, 1937-pattern webbing, with its deep ammunition pouches, had been introduced in 1937 to allow soldiers to carry the curved 30-round magazine for the Bren LMG. The issue rifle was the

These Royal Marines, their faces lined with fatigue and narrowing their eyes against the bright tropical sunshine, have just returned from a cross-border Claret operation into Indonesia. They are armed with 5.56mm Armalite rifles and have constructed 'belt orders' of ammunition pouches and water bottles around quick-release parachute cargo straps. (RM Museum)

reliable bolt-action No. 4 Rifle and Sten Mk V SMG. The webbing would remain in use into the late 1950s – blackened with boot polish to make it more water-resistant. Clothing was either navy blue or khaki battledress. The latter was a wool serge, two-piece garment consisting of a short jacket that buttoned to the trousers. The jacket had three pockets while the trousers had two, with a map pocket on the lower thigh.

However, in Malaya in the 1950s and later in north Borneo in the 1960s where men received the olive drab lightweight 1944-pattern webbing their weapons included the compact No. 5 jungle carbine, shotguns and the Australian Owen SMG. Initially there was no large pack available with 1944-pattern webbing, and some men carried a variety of packs including Bergen rucksacks and even 1937-pattern large packs. A 1944-pattern large pack was developed, but it was common to see the small pack with ponchos and other items strapped to the exterior. The first ponchos were made from heavy, rubberized cotton and men were keen to acquire the superior Australian nylon/PU ponchos that were lighter and stronger. Radios, such as the troop A41, were carried strapped to an aluminium pack frame – another piece of equipment developed as part of the 1944-pattern webbing. The clothing consisted of olive drab trousers and the 'shirt cellular'. They were as cool as such material could be in the humid tropics, but being made of cotton in thick jungle they quickly snagged on thorns, so at the end of a long patrol kit was often tattered and ripped. The boots were of rubber and canvas with lacing high up the leg in order to provide protection against leeches and other jungle parasites. By the end of the Confrontation in north Borneo (Brunei, Sarawak and Sabah) 1958-pattern webbing was being issued.

1958-pattern webbing was made of olive drab cotton webbing and retained the deep ammunition pouches first seen with the 1937-pattern, since the L4A4 7.62mm Bren was still in service. Marines were now equipped with the 7.62mm Belgian-designed FN self-loading rifle, universally known as the SLR. To complement it the belt-fed GPMG had entered service. The deep pouches of 1958-pattern webbing were useful not only for carrying the 20-round box magazines for the SLR and the 30-round magazines of the LMG, but also belts of GPMG ammunition. In addition to the 9mm Sterling SMG, the lightweight American 5.56mm Armalite rifle was carried. An unusual but effective way of ensuring that rain did not penetrate into the rifle and cause the working parts to rust was to put a Durex condom over the muzzle. Condoms were issued as prophylactics to troops in the Far East, in packages with a NATO stock number, but were used in this unorthodox way to keep watches waterproof as well as protecting rifles.

Personal load-carrying equipment could be modified, or older items (like the 1944-pattern water bottle and aluminium mug) retained following the issue of 1958-pattern webbing. Hot drinks could be prepared in the metal mug (in contrast to the black plastic 1958-pattern mug that replaced it); the trick was to stick a strip of 'pusser's tape' (black masking tape) along the rim

Men from the same group that has just returned from a cross-border Claret operation join their comrades, before the three-ton truck takes them back to showers, decent food and a debrief on their patrol. While some patrols would produce contacts and firefights, others might only gather intelligence. Both, however, would take the war to the enemy and ensure that cross-border raids were kept to a minimum. (RM Museum)

to prevent the drinker burning his lips. It looked (and probably was) insanitary, but it worked.

Where men had received stores and ammunition by parachute drop, the quick-release buckles from the harness that secured the loads were used to replace the buckles on their 1944-pattern belts. These quick-release buckles allowed belts to be easily adjusted, and eventually men adopted a 'belt order' without the supporting cross braces or yoke on the webbing. An item of equipment from this period that would remain in service for many years was the Golok. This 18-inch (45.7cm)-long machete was developed from a Burmese design with a tapered carbon steel blade. It was well balanced, and since the blade was made from relatively soft steel it would not chip; however, while it could be sharpened, it easily lost its edge after heavy chopping through bamboo. The hardwood handle was often bound with cord or twine for a better grip. Locally made machetes constructed from old car springs with distinctive wooden scabbards were popular with men serving in north Borneo.

One of the distinctive items of kit issued to Royal Marines serving in temperate climates were 'boots SV'(standing for 'soles Vibram'). These excellent ankle boots had a cleated rubber and leather sole that gave very good traction, and had been pioneered by No. 4 Commando in World War II.

In Korea 41 (Independent) Commando were issued with US M1943 combat clothing, webbing and US weapons; however, they retained their own boots. The US combat clothing was comfortable and practical and is the basis for almost all combat kit worn today.

In the bitter Korean winter the men wore issue underwear, long johns and string vests, angola shirts or a US equivalent, one or more woollen jerseys, battledress tops and trousers or US equivalents, combat jackets and trousers and pile-lined parkas with hoods. (A USMC parka as worn by 41 Commando is displayed in the Royal Marines Museum, Eastney, Portsmouth).

With Combat Dress the camouflaged Dennison smock originally introduced in October 1944, and often associated with airborne forces, was adopted for operations in more temperate climates, including Cyprus. It was a good medium-weight jacket with deep pockets and featured an effective camouflage pattern.

For operations in the Radfan, Aden and Kuwait Royal Marines wore khaki drill uniforms that in design were very similar to jungle uniforms. For patrols on the streets of Aden shorts were issued as part of No. 10 tropical drill dress. In the jungles and desert the beret was replaced by a soft-brimmed hat that in rain and sun rapidly became shapeless, but protected the head and ears. In the Yemen, suede combat boots with direct-moulded soles (DMS) were issued, while as far back as Korea web anklets, originally part

INTERNAL SECURITY IN NORTHERN IRELAND, EARLY 1970s
Body armour, padded leather gloves and patrol boots were among numerous items of equipment introduced during the Troubles in Northern Ireland. The Royal Marine shown here (1) wears olive drab trousers and a plain olive drab combat jacket, without webbing. On top of this jacket, he is wearing early M69 body armour (2, also shown in side view, 2a); the late M69 body armour (3), with rubberized patches to prevent rifles slipping off the shoulder during firing, is also shown as an inset. He is carrying an early, primitive green riot shield (4). Also shown is a Makralon SA12 riot shield (5); in more severe riots Marines would deploy these, although they favoured the fast-moving tactics of the snatch squad that would run in and arrest troublemakers in the crowd. The Marine's head is protected by a 1944-pattern Mk II helmet with SA12 visor (6); the later IS helmet with visor is also shown (7). The use of rubber bullets and CS gas to break up riots made the black S10 gas mask (8) an essential item of kit for the serving soldier. Another essential item for the Marine on crowd-control duties was the wooden baton (9).

of 1937-pattern equipment, had been replaced by khaki wool ankle puttees that gave better support.

In the late 1950s olive green combat clothing replaced battledress or cotton drill denims for operations in temperate climates. Designated No. 11 combat dress, it consisted of an olive green zippered jacket with four external pockets and trousers with three pockets and a map pocket. Made from close-weave sateen cotton fabric when first issued, it was water repellent. It was considerably superior to previous clothing and remained in service until the early 1970s. The original combat clothing reflected experience in Korea. The jackets had a 'poacher's pocket' at the back and the pockets at the bottom were elasticated. Knees and elbows were reinforced and the trousers that were lined had a two-way zip on the flies to ensure limited exposure to cold.

In the early 1970s the British land forces adopted a four-colour camouflaged design for combat clothing designated disruptive pattern material, or DPM. At the time many NATO armies still favoured olive drab and the US philosophy was that camouflaged clothing reflected a static passive mindset. DPM was, and is, a very effective design both in temperate and jungle environments and was adopted by the Royal Netherlands Armed Forces.

Initially the DPM combat clothing was produced as an exact copy of the earlier olive green combat kit. However, experience in Northern Ireland, where body armour or 'flak jackets' of increasing sophistication had been introduced in the 1970s, led to a new pattern of combat kit. The new jacket had a 'crotch flap' similar to the parachute smock and a pencil pocket on the sleeve – more accessible when the man was wearing a flak jacket.

The long conflict in Ulster that began in 1969 would see 13 Royal Marines killed in action in the Province, in addition to 11 men from the Royal Marines School of Music killed in an IRA bomb attack at Deal in England in 1989. During the course of the conflict a range of protective clothing was introduced – the most distinctive was the fragmentation vest bought from the United States. It was upgraded and improved as the conflict dragged on and the weapons employed by the IRA became more lethal. By the close of the conflict the body armour included ceramic plates as well as non-slip areas on the shoulder against which a rifle but could be rested. Other kit and clothing included padded Northern Ireland (NI) gloves (elasticated leather gloves that protected the hands when taking up fire positions in urban areas) and the NI boot – lighter and higher-lacing, to give better ankle support.

The Mk III steel helmet – new on D-Day in 1944 – had soldiered on for many years. It was fitted with a Makralon visor as protection against rocks and other missiles, but in 1975 a new internal security helmet, similar in design to those worn by motorcyclists was introduced.

In the Falklands, although some of the Guards regiments wore the Mk III helmet, the Marines favoured their berets – practical, light and not top heavy, and a reflection of their *esprit de corps*.

Where experienced tailors were available (and they were at most barracks), Marines would have their combat kit modified. The most popular and practical amendment was the addition of a deep map pocket in the jacket that was accessible via a vertical zip. Eventually with the 'Combat Soldier 90' range of kit, many of these modifications would become standard.

The green polycotton TMLs – trousers, men's, lightweight – were a popular item of clothing often worn with the DPM combat smock. TMLs dried quickly and were light and hard wearing. However, polycotton, as the Royal Navy discovered in the Falklands, would melt and stick to the wearer's skin if subjected to high temperatures at close range.

Operations in Northern Ireland, an area that enjoys a regular supply of rain, saw the issue of PU-coated nylon clothing. This was either in DPM or reversible dark green and white for the new NATO role of arctic warfare assigned to the Royal Marines covering Norway on the northern flank. For this they developed new skills and equipment – Mountain and Arctic Warfare (M&AW). Their superior Arctic Smock – a DPM hooded smock with deep pockets and large buttons – together with trousers with deep pockets would prove ideal in the bleak weather of the Falklands.

As part of this clothing Marines were issued with a quilted liner and trousers, known as a Mao suit after the quilted uniforms worn by the Chinese Army. This gave excellent insulation. In a snow hole or tent, boots could be discarded and replaced with green quilted duvet boots. Other footwear included 'moon boots', very large insulated overboots made from canvas that can be worn over boots ski/march or duvet boots. Toecaps and rubber slipovers were worn when a man was static for a period, since heat loss is faster from the toe area. Experienced Royal Marines used the cardboard from ration packs when they were standing on sentry duty to further insulate their feet from the snow. The discipline of maintaining a smart turn-out and appearance paid off in the Arctic where clean clothing gave better insulation than soiled. Men carried six pairs of socks as part of their basic load.

In Norway Marines would wear white snow camouflage smocks and trousers, made from lightweight nylon. These not only concealed the men but shrugged off snow that would stick to cotton fabric. Olive green and later DPM cold weather caps with earflaps were a popular item and the Commandos had green Gore-Tex gaiters that kept snow out of the top of their boots ski/march. These versatile square-fronted boots would accept the ski bindings to allow men to travel cross-country on skis. Padded DPM mitts with a 'trigger finger' gave protection to the hands – but some men would suffer cold injuries to their index fingers when skiing, forgetting that their trigger finger was not insulated. Where mitts were not practical, white knitted fingerless gloves allowed a degree of dexterity, but also protection to the back of the hands where the blood vessels are close to the surface. One item of arctic warfare kit that was, and remains, very popular was the insulated vacuum flask. A conventional steel thermos with cup, it had a two-part cell foam insulated outer cover with strap that allowed the owner to sling it over his shoulder when not in use.

It takes time and fuel to melt snow for water, so a ready supply of hot liquid is always welcome. One technique used by experienced

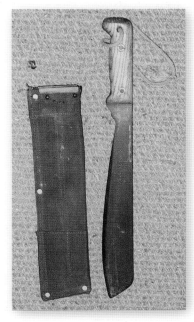

The 1944-pattern Golok machete. This well-balanced tool came in a reinforced dark green webbing sheath that could be clipped to the 1944-pattern belt. For silent movement gardening secateurs were used to snip through the jungle. (Author's collection)

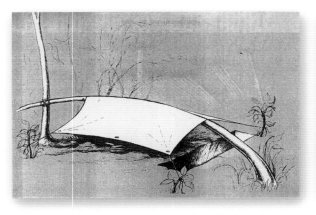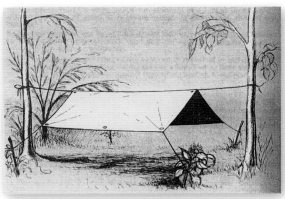

Away from company bases on the Indonesian border, men on patrol would construct overnight temporary shelters using ponchos and para cord. Universally known as *bashas*, these would give cover against the heavy tropical rain. They remain a standard operation procedure for the British Army and Royal Marines. The low-slung *basha* (left) is sited by a slit trench, the 'stand-to' position for the marine; the higher *basha* (right) would be set up in a more secure area. (Author's collection)

Marines was to make enough hot water to fill a water bottle. With the cap firmly screwed in place this served as a hot water bottle in the sleeping bag and by dawn was still liquid and ready to be boiled up for a 'wet'.

The robust Wilkinson MOD survival knife was issued to men training in Norway. This 12.25-inch (31.1cm)-long, 1 lb 2oz (0.5kg) knife was basically a steel bar with a sharpened edge at one end and wooden grips riveted on at the other. It was a good chopping tool, but after extended periods of work the wooden grips were uncomfortable.

When the Royal Marines spot a practical piece of kit, they are quick to adopt it. The zippered cotton 'Norwegian shirt' that has now become an almost standard item of clothing was adopted from the Norwegian Armed Forces, along with the excellent sweaters and robust daypacks.

Among the threats posed by the Soviet Union and Warsaw Pact was the use of battlefield chemical weapons. For this men were issued with nuclear-biological-chemical (NBC) suits, sometimes known as 'Noddy suits'. These consisted of a hooded pullover smock and loose trousers made from a material impregnated with charcoal. Feet were covered with overboots, and hands with tough, rubber gloves and cotton liners. The face and lungs were protected by an S6 respirator (later S10) that had a detachable filter. With the use of CS gas to break up riots in Northern Ireland, the S6 became a piece of equipment with which men were very familiar. In conventional war, when faced by a lethal gas attack the respirator had to be donned in nine seconds – the cry being 'Mask in nine!' In warm weather the NBC clothing was hot and uncomfortable, and in the arctic winter in Norway it was even more trying. However, during the campaign in the Falklands some Marines found that the NBC overboots offered useful waterproof protection in the sodden marsh and heather of the islands.

BELIEF AND BELONGING

Long before they took on the Commando role, after World War II the Corps of Royal Marines had a proud and distinguished history as 'sea soldiers', something that has formed the core of their identity. Emphasizing this, the 1968 booklet *Notes for Recruits* issued at Deal states:

> In joining the Royal Marines you inherit certain traditions of which we are very proud and jealous. We have a record of service all over the world which goes back for three centuries. A Corps which has lived for so long and has fought in so many wars ashore and afloat is bound to have acquired traditions

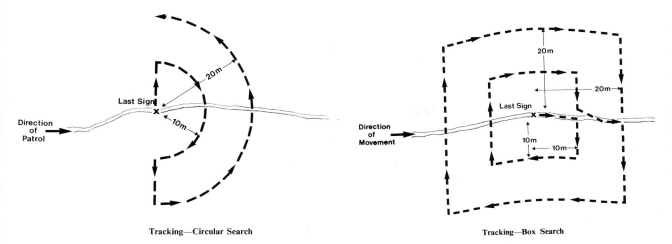

Tracking—Circular Search

Tracking—Box Search

Tracking was a skill that could be learned with practice. 'Jungle lore' classes would teach soldiers and Royal Marines how to live in the jungle and also how to track. If contact was lost there were techniques for picking up the trail again – the circular search or the box search. These might locate discarded food wrappers, broken branches or the prints of combat boots in soft ground. (Author's collection)

of duty and comradeship which are worth treasuring. These are traditions in which you are now entitled to share. But your share imposes certain responsibilities. See to it that in everything you set your hand to, throughout your service in the Corps, you do your level best.

In training, recruits were taught about the origins and history of the Corps and the way the distinctive cap badge evolved. On 24 July 1704 six battalions of Marines captured Gibraltar in three days, and then held it for an eight-month siege. On 7 June 1761 at the Battle of Belle Isle, Marines were awarded the 'laurel wreath' to the cap badge for their part in storming and capturing the island off the coast of France. Both these actions are Corps memorable dates. On these dates the guard presents arms and a fanfare is sounded on silver bugles.

On 17 June 1775 at the Battle of Bunker Hill in the American War of Independence the Motto 'Per Mare per Terram' was worn on the caps of Marines in action against the colonials. On 29 April 1802 King George III directed that the Corps of Marines be styled 'The Royal Marines'. With this title the lion and crown were added to the cap badge and facings changed from white to blue. On 26 September 1827 the Duke of Clarence, later William IV, presented colours to all divisions and announced that, because of the great number of battle honours to be considered (109), King George IV had directed that the 'globe encircled with laurels' should be the badge of the Royal Marines. He also ordered that his own cypher 'GR IV' was always to be borne on the colours. 'Gibraltar' was to be borne as the sole battle honour.

Between 1854 and 1855 the first three Royal Marine Victoria Crosses (Britain and the Commonwealth's highest award for bravery) were won, two in the Crimea and one in the Baltic during the war with imperial Russia that became known as the Crimean War.

On 7 March 1918 King George V inspected the Depot Royal Marines, Deal and decreed that the Senior Squad should be titled 'The King's Squad'. The best recruit of the squad became 'The King's Badgeman' and wears the Royal Cypher 'GR V' surrounded by a laurel wreath on his left shoulder.

St George's Day is the Corps' remembrance day, and commemorates the 23 April 1918 raid on the German-held port of Zeebrugge in Belgium. The 4th Battalion Royal Marines formed part of the landing force for the raid and two VCs were awarded to Marines.

The amalgamation of the Royal Marine Artillery (RMA) and Royal Marine Light Infantry (RMLI) on 22 June 1923 produced the distinctive colours of the Royal Marines, the red and blue from the RMA and green from the RMLI bugle cords. The yellow dates back to the facings on the coats worn by the original formation in 1664.

Among the Corps' memorable dates from World War II are the 19 August 1942 raid on Dieppe by 40 Commando; 9 September 1943 landings at Salerno, Italy by 41 Commando; 6 June 1944 D-Day landings, in which 18,000 Royal Marines took part; 7 June 1944 capture of Port-en-Bessin; 11 June 1944 attack on Le Hamel and Rots by 46 Commando; 1 November 1944 Battle of Walcheren, where 41, 47 and 48 Royal Marine Commandos landed and fought for the strategic island covering the approaches to the Belgian port of Antwerp; 23 January 1945 attack on Montforterbeek, Holland by 45 Commando; 31 January 1945 landings at Kangaw, Burma by 42 and 44 Commando; and 2 April 1945 Battle of Lake Comacchio, involving 43 Commando. It was at Lake Comacchio that the Commandos won their first (posthumous) VC when, armed with a Bren LMG, Corporal Tom Hunter put in a bold one-man attack against German machine guns that were holding up the advance.

The Commando role, inherited by the Royal Marines in the aftermath of World War II, has ensured that they have remained a distinctive organization within the British armed forces. On 7 July 1961, the words of HRH the Duke of Edinburgh, Captain General of the Corps, when presenting colours to 41 Commando, summed up what made Royal Marine Commandos a special force then and to this day:

> The whole value and essence of Royal Marine Commandos was that they could undertake any kind of military or security job under any conditions anywhere in the world at a moment's notice. This demands a very high standard of training without let-up, and if anyone is inclined to question this let them consider recent events in the Persian Gulf. The Commando Ship was on the spot and the Commando ashore without a moment's delay. This is what we expect, but is not achieved without constant preparation and attention to detail, or without constant improvement of methods or techniques. The Corps of Royal Marines have a splendid tradition of courage and smartness going back many years, but you cannot live on traditions. You have got to be ready to cope with the hard and practical realities of today, with an eye to what might happen in the future. If you succeed in keeping up to date the tradition goes on – if you fail you go out of business. The Royal Marines show no signs

D **ROYAL MARINES TAKE A BREAK DURING THE WORK-UP TRAINING PHASE FOR OPERATIONS IN NORTH BORNEO IN THE INDONESIAN CONFRONTATION OF THE MID-1960s**
The Marines have stopped to make a *basha* shelter for the night. All are dishevelled and unkempt, their uniforms heavily creased, sweat-soaked and dirty after a day in the jungle. Their trousers and shirts are the tropical issue, in olive drab cotton, and they wear jungle hats and a net scarf around their chest and necks to absorb sweat.

The lessons of the war in Burma led to the development of the excellent 1944-pattern webbing, as well as olive green uniforms and high-lacing jungle boots – the latter designed to prevent leeches and other jungle parasites penetrating to the wearer's skin. The 1944-pattern webbing included a small pack with a waterproof lining and an aluminium water battle and mug – the latter excellent for making a quick 'wet'. Among the weapons shown are a machete, a Sterling submachine gun (L2A3, Mark 4) and an L1A1 self-loading rifle and the re-barrelled Bren Light Machine Gun configured to fire 7.62mm ammunition.

of failing, and I am quite sure that 41 Commando will keep up the tradition of worrying about the next battle and not about the last.

A few words of explanation regarding the above: in 1961 both 42 and 45 Commandos were absent, having taken up positions in Kuwait to protect the oil rich state from an Iraqi armoured threat. The Commandos selected the Mutla Ridge on the road north out of Kuwait City – the ridge would later become notorious as a 'killing ground' in the 1991 Gulf War.

Today the Royal Marines have a formula for the values that prepare them for the next battle. These were a 'given' for many years, but formalized as the Corps ethos they incorporate the individual Commando spirit with collective group values. These are courage, unity, determination, adaptability, unselfishness, humility, cheerfulness, professional standards, fortitude and finally a uniquely Commando humour.

LIFE ON CAMPAIGN

The defining feature of the operations undertaken by Royal Marines from Korea to the Falklands was the resilience and initiative displayed by the men involved. The jungles of Malaya and Borneo were strange, silent places, while the deserts of the Radfan and Kuwait and the streets of Aden were hot and harsh. The fields of South Armagh and streets of Belfast and Londonderry looked disturbingly familiar to men from the mainland – the shops were the same and so was the terrain. In all these theatres operational responsibility was often passed down the line to section commanders. In Northern Ireland it sometimes fell to these young men to take quick decisions that might become a major news story by the evening and have politically significant outcomes.

Whether it was the bitter cold of Korea, the humid heat of the jungle or the desiccating heat of the desert, Royal Marines had and have a close-knit loyalty. This ensures that they watch out for the symptoms of dehydration or frostbite among their section and between Marines on an individual basis.

In Korea the men of 41 (Independent) Commando retained their berets, but were otherwise equipped and armed by the US Marine Corps. In his authorized history *The Royal Marines 1919–1980* James D. Ladd notes:

American kit they found to be generally satisfactory, even if the warm sleeping bags could not be used in forward areas, as they were 'difficult to get out of in a hurry' and that half full water bottles made too much noise for silent patrols at night.

The Commandos were issued with a range of US weapons, including the Browning Automatic Rifle (BAR), the M1 and M3 carbine, M191A1 Colt pistol, M1 Garand and 3.5-inch M20 rocket launcher or 'super bazooka'. However, like their American comrades they were unimpressed by the M3 'Grease Gun' and where possible attempted to get hold of the M1926 Thompson SMG. The World War II veterans were familiar with

The 1944-pattern large pack would be required to accommodate all that a man might need on two-day patrol inside hostile territory, where the luxury of aerial resupply was not available.
Pocket A: 1. Face veil. 2. Insect repellent. 3. Foot powder. 4. Matches. 5. Water sterilizing kit. 6. Rifle cleaning kit. 7. String or parachute rigging line cord.
Pocket B: 1. One mess tin. 2. Spoon or fork. 3. Snack, tea, sugar, milk. 4. Hexamine solid fuel cooker.
Main compartment: 1. Two 24-hour ration packs. 2. Spare socks. 3. Large waterproof bag. 4. Medical pack (section commander). 5. Gym shoes strapped outside.
Pocket D: 1. Poncho. 2. 120 feet of rope cord or cod line (section commander).
In addition to these basics a patrol commander might carry binoculars, maps and a protractor, with his second-in-command carrying a similar extra load. Where patrol dogs were accompanying the operation, the dog handler would take two 1 lb cans of dog meat. (Author's collection)

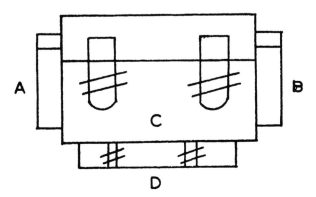

A Royal Marine sniper with the equipment required for operations in the Arctic. It includes the quilted Mao suit, underclothes, skis, snow shoes, 'duvet boots', thermos flask and insulated carrier, arctic sleeping bag and mat. The Royal Marines developed a layered clothing system that trapped warm air against the body, which could be adjusted if the temperature rose. (Commando Forces News Team)

some of these weapons and were said to be happy with the M1 Garand, but found the BAR a poor substitute for the Bren. Perhaps the weapon with which they were least happy was the M1 fragmentation grenade, since there were many blinds. Therefore, the only British weapon used by 41 Commando was the No. 36 Grenade.

The operation in Cyprus against the Greek nationalist EOKA terrorist organization in the 1950s saw the Commandos taking on a counter-insurgency role. Patrols combed the mountains searching for hideouts where the EOKA operatives had concealed themselves; the Royal Marines employed dogs in these operations. Faced by rioting crowds in the cities of Cyprus the Commandos pioneered the concept of the 'snatch squad', later used in Northern Ireland. Helmets, batons and riot shields were set aside and squads in light order wearing gym shoes pursued and arrested rioters.

In the Emergency in Malaya in the 1950s, and later the Confrontation in North Borneo in the 1960s, Royal Marines learned or re-learned jungle skills. The Malayan campaign between 1950 and 1960 was fought against Communist Terrorists (CTs) of the Malayan Races Liberation Army. This Chinese-driven insurgent organization was intent on establishing a

A Volvo Bv202E tracked over-snow vehicle prepares to tow a Royal Marine Commando patrol during training in Norway in the 1980s. As part of their arctic warfare training Royal Marines learned cross-country skiing carrying a weapon and Bergen rucksack. It was this training that would pay off during the Falklands campaign in 1982. (Commando Forces News Team)

communist government in the former British colony, which was soon to be given its independence. It was a campaign fought by a mixture of tough World War II Marine veterans of Normandy and Walcheren, and young men called up for National Service. Sir John Harding Chief of the Imperial General Staff would write of the work of the Royal Marines during the Emergency: 'It is a record of hard work, devotion to duty and good comradeship of which the Royal Marines have every reason to be proud.'

The Confrontation was a period between 1962 and 1966 in which cross-border raids into Sarawak and Sabah by insurgents and later regular Indonesian Army soldiers were used by President Achmad Sukarno in an attempt to destabilize the newly formed Malaysian Federation and establish a wider nationalist and communist empire. The UK had treaty obligations with Malaysia, and so British soldiers and Royal Marines found themselves in a challenging conflict that required them to relearn the skills of jungle warfare from the Emergency.

Smell, whether in the jungles of Malaya or Borneo, carries over long distances so Marines stopped washing and refrained from smoking – which seems obvious, but they even gave up sucking the boiled sweets that came in ration packs as they produced distinctive odours.

In the Confrontation Commandos were deployed along the border with Indonesia to halt incursions into Sarawak and Sabah. They developed good relations with the villagers who lived along the border, offering low-level medical assistance and other help. In the words of the operational doctrine

A ROYAL MARINE MACHINE GUNNER IN THE RADFAN MOUNTAINS, ADEN, 1965
This illustration shows a Royal Marine machine gunner, carrying the General Purpose Machine Gun (GPMG), in the Radfan Mountains of Aden (now Yemen), c. 1965. He is standing in front of a rock *sangar* shelter, which his fellow Royal Marines have helped to build. Dressed in a cotton khaki drill shirt and trousers and wearing a battered hot weather hat, he sports 1958-pattern webbing in sun-faded olive drab with ammunition pouches attached, and the boots SV. Slung across his chest is a belt of GPMG linked ammunition; the black metal links are positioned to face outwards to reduce the glint from the brass cartridge cases. Between 1964 and 1965 the harsh conditions in the limestone mountains of the Radfan quickly wore out boots and clothing. However, here troops were on 'wartime accounting', and there was no requirement for complex paperwork to replace worn-out kit.

A Royal Marine Rigid Raider assault boat speeds up a Norwegian fjord during training in 1984. The Rigid Raider is a robust craft, and its deeply raked bow allows it to be driven onto shallow shelving beaches. The coxswain is well protected against wind chill and is an wearing immersion suit. (45 Commando Condor Photographic Section)

A Fleet Air Arm Wessex helicopter kicks up fine snow as men disembark. In the foreground a Royal Marine wears an arctic smock and white nylon camouflage trousers – probably to allow loose snow to be brushed off easily. The dark green outline of his 1958-pattern webbing has been broken up with strips of white tape and the SLR is similarly camouflaged. (Commando Forces News Team PO, P. Holdgate)

of the time they 'won the hearts and minds' of the population. This was invaluable, since in the 1960s they were able to build up an intelligence picture of the border areas.

Commandos became expert at roping down from hovering helicopters into small areas of open ground in remote areas of the jungle and then moving on foot to set up ambushes to catch cross-border raiders. Ambushes required patience and the ability to endure jungle pests in complete silence. Food and water would be eked out and movement kept to a minimum. While the nights could be surprisingly cold and damp, the heat of the day could make the most alert man start to drift off.

The Commandos developed the techniques for tracking and routines for setting up temporary overnight bases deep in the jungle. The name for shelters made from ponchos stretched between trees was the native word *basha*, a term that remained in use after the 1960s for any overnight shelter. Skilled Dyak and Iban trackers from Sarawak had served in Malaya in the 1950s as trackers, and now their sons brought these skills to Borneo during the Confrontation.

Lightweight jungle rations developed for the Confrontation produced about 2,000 calories per day. Like all British '24-hour rations' it consisted of a breakfast meal, snack, main meal and sundries. Breakfast was an oatmeal block, instant coffee, sugar and milk powder. The snack consisted of tinned processed cheese (known by all as 'cheese possessed') biscuits and chocolate. The main meal was dried

Dressed in his white nylon snow camouflage clothing, a Royal Marine of 45 Commando moves cross-country on his skis. His rifle is slung over his pack on a two-point sling, and white tape has been used to break the outline of his green map case. He wears an ear-flapped fur pile cap. (45 Commando Condor Photographic Section)

beef granules, potato mash powder, instant tea (which was very unpopular) sugar and milk powder. The sundries packet contained chewing gum, salt, six sheets of toilet paper and a book of matches. The entire ration weighed slightly over 15 ounces. The British Army training manual *Land Operations, Volume V: Operational Techniques, Part 2: Jungle B: Jungle Skills and Drills* notes: 'If this ration is consumed continuously for periods in excess of 10 days a period of recuperative feeding must be undertaken.'

New one-man 24-hour ration packs were introduced later with different menus, which consisted of a mixture of tinned and packaged food and beverages. Men became adept at combining these with local ingredients. Surprisingly, the salty and unpopular 'cheese possessed' lingered on. Though bulky, the advantages of tinned rations were that they did not require water for reconstitution, and the water in a mess tin used to heat up tins could be put to good use for a 'wet'.

A patrol of 45 Commando in the streets of Belfast in 1977. The Marine wears a camouflaged combat smock and green 'trousers, men's, lightweight' (TMLs), and padded-leather Northern Ireland gloves designed to protect the hands when taking up fire positions on hard surfaces. He has a fragmentation vest with non-slip patches on the shoulders. His SLR is secured to his wrist. In the background is an armoured 0.75-ton Land Rover. (RM Museum, 7/20/16 (172))

A Marine armed with an SLR and baton-round gun pauses during a patrol in the Turf Lodge district of Belfast in 1977. The skills honed in the streets of Northern Ireland were used in Northern Iraq in 1991 when, as part of Operation *Provide Comfort*, Royal Marines patrolled in the Kurdish town of Zakho, pushing back Iraqi forces that were threatening the town. (RM Museum, 7/20/16 (33))

Besides the 24-hour ration, other rations were issued as ten-man packs that could be cooked centrally by a section. Ideally food was prepared in field kitchens from a mixture of fresh and tinned ingredients and brought forward to troops in the line in insulated containers. Tea or soup came in a green vacuum flask about two feet long and known as a 'tea bomb'. Stews were cooked in a dixie – a deep cooking pot with a shallow lid that could be used as a frying pan. The dixie that was still in service in the 1970s dated back to the beginning of the 20th century. Another item of collective cooking were the 'hay boxes' insulated with hot water to keep the inner food container hot. Training with the Norwegian Army introduced the Royal Marines and British Army to insulated food containers that became universally known as 'Norwegian containers'. These were lighter and used plastics and alloys in their construction – their shape also allowed them to be strapped to packframes and carried cross-country. Whilst training in Norway, the Commandos learned that their arctic warfare rations contained about 5,000 calories and that it was essential that they eat them at intervals during the day. What came as a surprise was that they needed fluids in the same volumes as if they had been training in hot weather. Tea and coffee were supplemented by hot soups and cocoa to ensure the correct fluid intake and add extra calories to the meals.

Two items of equipment were crucial for operations in the jungle: a Millbank bag, and water sterilizing tablets. The former was a close-weave cotton filtering bag open at one end that tapered to a point. When water was poured into the bag it trapped any large organic matter but allowed the clear liquid to drip off the pointed end into a water bottle. The Marine then

An Argentine officer in an amphibious vehicle shouts at the Special Forces who have taken prisoner the men of Naval Party 8901 at Port Stanley on 2 April 1982 during the invasion of the Falkland Islands. He is demanding that the humiliating treatment should stop and the men be allowed to stand up again. However, the anger these pictures provoked made British public opinion support the liberation of the Falklands. (Private collection)

added one of two water-sterilizing tablets; this killed bacteria but left the water tasting strongly of chemicals. The second tablet was added just before drinking to remove this taste.

The jungle is home to numerous tropical diseases. Mosquito-borne diseases include malaria, dengue fever, filariasis and virus encephalitis. In addition there is scrub typhus, lepospirosos and bowel diseases, skin diseases and, the most common hazard of all, heatstroke and heat exhaustion. Men learned to watch their battle partner closely for signs of any of these ailments to ensure prompt treatment and evacuation.

While the jungle abounded with diseases the Arctic posed three major threats: frost nip, superficial frostbite and deep frostbite. To these could be added snow blindness, sunburn, and carbon monoxide poisoning or fume irritation from poorly ventilated tents, and constipation. As per spotting heat exhaustion, noting the symptoms of frost nip and bite were on a buddy basis in which each pair of Marines would watch for the symptoms in their buddy and take remedial action.

Before they were deployed to the Far East, Marines fought in the Radfan, the mountainous regions north of the port of Aden at the southern end of the Red Sea. In Aden the promise of independence to this British colony had produced anti-British attacks by nationalists, who hoped to accelerate this departure and be seen as the force that had ejected the colonial power. The Royal Marines along with British Army infantry battalions were given the invidious task of keeping the peace up to withdrawal. Street patrols covered the town checking for weapons and contraband.

The operations in the Radfan were tough, but more in keeping with the role for which the Commandos had been trained. It was impossible to dig in the hard terrain and the Marines became adept at building *sangars*, the simple defences constructed from rocks and small boulders. The word *sangar* survived in use into operations in Northern Ireland describing the positions constructed from sandbags and corrugated galvanized iron (CGI), also known as 'wriggly tin'.

With the barrel of an 81mm mortar strapped to his packframe, a Royal Marine of 40 Commando yomps up the road up the volcanic mountain that forms Ascension Island during training on 10 May 1982. For the Task Force, the break at the island was an opportunity to test weapons, repack kit and prepare for the final leg south and landings on the Falklands. (RM Museum, 7/20/18 (16))

1968 is a unique year in British post-war military history: no soldier was killed on active service in the world. However, this was soon to change dramatically when the British Army was deployed to Northern Ireland in June 1969. For the Royal Marines 'the Troubles' would last until 2004 with Commandos deploying on regular roulement tours.

The veterans of 45 Commando had seen some of this before – riots, snipers and ambushes – when they had patrolled the streets of Aden town in 1967. However, Northern Ireland would see an increased threat from larger improvised explosive devices, or IEDs – and future veterans of Iraq and Afghanistan would encounter this threat in the 21st century. The tactics to defeat them were a mixture of intelligence, common sense and, as always in war, luck. On patrol, men learned not to take the obvious route, even if it was the easiest. Command wires leading to buried charges might be spotted because the soil had been disturbed, or intelligence might lead to a raid on a suspect house where bomb-making materials might be hidden. The covert observation post would play an important part in this intelligence-gathering campaign; concealed in derelict buildings or in natural cover on the border, Marines would watch the movement of vehicles and men and women, logging this and helping to build up a bigger picture. Night observation devices, which had once been bulky and cumbersome, shrank in size and became a valuable aid to patrolling and observation. As the Troubles morphed over the years, the Royal Ulster Constabulary (RUC) took primacy, with the Army and Royal Marines providing the back-up and if necessary the fire power.

The Provisional IRA, which had started its IED attacks with home-made explosives based on agricultural fertilizer, began to receive factory-made Semtex plastic explosive via Libya, and upped its operations. Timers, reliable detonators and the cruel and inventive mind of the bomb makers made these devices a constant threat. From sympathizers in the United States came funds and sophisticated small arms, such as the Armalite and M60 machine gun and later even .50-inch, long-range sniping rifles. This meant that no tour by a Commando was ever the same.

For the Marines Northern Ireland would see them housed in company bases, either on the border or in disused factories and mills in cities such as Belfast and Londonderry. For Marines on their first tour in the province, the overwhelming and disturbing experience was that it closely resembled home – the shops in the streets were the same, buildings similar and the signs in English. The very normality made it hard sometimes to realize that it concealed a hidden threat. The girl with a baby in a pram might be carrying a stripped-down rifle; the car parked by the road might conceal a bomb; the child sitting on a wall watching the patrol walk by might be a 'dicker' noting their numbers and direction and reporting back to the adults the movements of the Marines and their accompanying RUC officer. Adjusting to normality in England, Wales or Scotland while on leave was sometimes very hard.

The 1982 attack by the Argentine military junta on the small and passionately pro-British Falkland Islands in the South Atlantic was unexpected. However, it saw the rapid assembly in the United Kingdom of a formidable task force that included two elites: 40, 42, and 45 Commandos, and the 2nd and 3rd Battalions The Parachute Regiment. The challenge was to ship them thousands of miles south and land them safely on the islands. As the Task Force, comprising warships and the rapidly converted luxury liner SS *Canberrai* serving as a troop ship, moved south one British commentator rashly predicted that an attack on the Falklands 'would leave us up to necks in dead Marines'.

There were casualties during the landings, but mercifully losses in the campaign were fewer than predicted. The ability of British troops to live in the field and their professionalism paid off against a large but poorly trained Argentine force. One minor problem that had not been anticipated was that the high calorie arctic rations issued to the men had been developed for snowy Norway, where melted snow could be used to re-hydrate the food. Not only was there no real snow in the Antarctic winter of 1982, but there was the added threat that the numerous sheep on the island had infected the streams with a liver fluke. For six weeks one unfortunate Marine drew nothing but Arctic rations in which the main meal was Chicken Supreme. In the closing stages of the war, the Royal Marines of Naval Party 8901, who had been outnumbered and forced to surrender to the invading Argentine forces on 2 April 1982, returned to the island to raise the island's flag at Government House on 15 June.

On board a Landing Ship Dock off San Carlos Water, East Falkland; men wait with faces camouflaged and kit packed to board the landing craft that will float out when the dock is flooded. Until 1982 there were sceptics who believed that amphibious warfare skills perfected by the Royal Marines were no longer relevant. As a result, the crew of HMS *Intrepid* had been paid off and the ship mothballed. (RM Museum)

THE MARINE IN BATTLE

Many of the tactics and techniques employed by Royal Marine Commandos were identical to those of British Army infantry battalions. The basic building block was, and remains, the section of about ten men commanded by a corporal. Three sections together with a troop HQ made up a troop, and there were three troops to each company.

In attack the aim was to put down sufficient fire so that the enemy would 'keep their heads down', allowing men to close with the position and capture it. Fire and movement has been likened by instructors in World War II to a parrot walking up the bars of its cage – it always has one claw holding onto a bar when the other is reaching for a new grip. So too the fire section or Bren (later the GPMG) group would put down a heavy volume of fire as the assault section manoeuvred into position. Closing the range with the enemy would ensure that fire was more accurate and effective and that the final assault was delivered over a short distance.

Now ashore on East Falkland, the Intelligence Section of 3 Commando Brigade prepares to move following an air attack against their position. The Bv 202 has a 9 x 9ft tent fitted to extend the space in the back of the vehicle. Packs with sleeping bags are strapped to the side of the vehicle, and additional stores have been strapped to the roof. (3 Commando Brigade Photographic Section)

In addition to the section and troop automatic weapons, the troop would have a 2-inch mortar that could fire smoke to blind the enemy and screen the assaulting sections. The handy mortar, which could be carried by one man slung over his shoulder, had a maximum range of 500 yards (456m) and fired a 2.25 lb (1.02kg) HE bomb, which on hard ground could cause casualties up to164 yards (150m) from the point of impact. Other ammunition included smoke and illumination rounds. In the 1950s and 1960s Commandos had the American 3.5-inch rocket launcher until it was replaced by the Swedish-designed 84mm Carl Gustav recoilless rifle. The Carl Gustav had been used against suspect car bombs in Northern Ireland, but was first fired in anger in the Falklands by men of Naval Party 8901 on the night of 1/2 April 1982 at an Argentine LVTP-7 amphibious vehicle driving into Stanley on the day of the invasion. However, it was on 2 April in South Georgia that a detachment commanded by Lieutenant Keith Mills engaged the Argentine corvette *Guerrico* with a Carl Gustav and 66mm LAWs. Fire from these weapons damaged the warship and forced it to withdraw. Subsequently the Carl Gustav was used against Argentine bunkers following the landings at San Carlos.

The machine guns in the support company were initially the long-serving Vickers MMG. On 11 December 1962 at the beginning of the Indonesian Confrontation a hostage rescue operation took place at Limbang in Sarawak in north Borneo. Men of L Company, 42 Commando commanded by Captain Jeremy Moore saw a section of MMGs giving supporting fire from a commandeered civilian lighter in the Limbang River. When the machine-gun section commander Quartermaster Sergeant Cyril Quoins asked the Royal Navy officer commanding the lighter if he could pull the craft out of line to give the guns a better field of fire the officer's reply was pure history: 'Sergeant-major, Nelson would have loved you!' The lighter moved into a more exposed position and the MMGs blasted the buildings held by the rebels, before the Commandos raced ashore to rescue the hostages. Other weapons that formed part of the armoury of the support company included the 3-inch and later the 81mm mortar, the GPMG and the Milan anti-tank guided weapon (ATGW).

105mm light guns fire on East Falkland. The Commando gunners were invaluable in their ability to provide quick and accurate fire in support of the attacks against the defended features that blocked the western approaches to the island's capital Stanley. The guns could fire a 35 lb (15.8kg) shell a distance of 10 miles (16km). (Media Ops, Land Command)

While medium mortars were very effective in the defence with the mortar fire controller registering defensive fire (DF) tasks they could also be used to neutralize enemy positions in the attacks. Mortars could fire smoke, HE or illumination rounds and before effective night vision equipment had been developed 'White Light'; the light from mortar bomb flares descending on their parachutes was the only real source of illumination on the battlefield.

While a 2-inch mortar crew could observe the explosions from their bombs, the 3-inch and 81mm worked with a mortar fire controller (MFC). Linked by radio to the mortars he would correct their fire on DFs. In a defensive position the DF final protective fire (FPF) was the barbed wire to the immediate front of the company position. If the DF FPF (also DF SOS) was called, the mortar crews knew that the fighting had become very serious.

ROYAL MARINE COMMANDO CORPORAL AT SUEZ, NOVEMBER 1956
This Royal Marine corporal (his rank indicated by the double-chevron on the upper sleeves) wears the four-pocket camouflaged Dennison smock (1) issued to airborne forces and the Royal Marines. Beneath this he wears khaki battledress (2), which was worn both as combat clothing and also as a parade uniform. Although it was warm in cold weather, the battledress fabric absorbed water, the short blouse was impractical, and the map pocket on the trousers (3) was not large enough to accommodate maps and notebooks. He wears the excellent 'boots SV' (soles vulcanized/Vibram, 4) which are topped off with blackened 1937-pattern webbing anklets (5), also shown in inset view. Details of his polish-blackened 1937-pattern webbing are shown as insets, including his belt (6), ammunition pouch (7) and small pack (8), which contains his folded groundsheet cape. Correctly fitted, the webbing was practical and well balanced. His beret (9) which features the famous metal cap badge, is the standard Royal Marine dark green, and he is armed with the Lee-Enfield No. 4 rifle (10). Also shown are No. 36 hand grenades (11) and the No. 80 white phosphorus grenade (12)

8

9

6

5

7

1

10

3

2

5

4

11

12

WP

REN NO. BOM
DY 9/55
LOT 10

To reach targets at an extreme range the barrel of the mortar was adjusted to a shallow angle and the maximum number of charge increments possible were clipped to the tail. Some 3-inch mortar veterans claimed that adding a mug of petrol down the barrel before a bomb was loaded would also extend the maximum range of 2,750 yards (2,516m). However, this may merely be a good yarn, since hot mortar barrels and petrol are a hazardous combination. The HE and smoke bombs weighed 10 lb (4.54kg), and an experienced crew could fire 15 bombs a minute. The mortar could be broken down into three loads – base plate 37 lb (16.7kg), barrel 42 lb (19kg) and bipod and sights 45 lb (20.4kg) – and carried by men or mules to remote locations.

During the Falklands campaign a 42 Commando 81mm mortar crew was photographed on Mount Kent hunched in the rain clad in their reversible green and white PU Arctic waterproofs. They were about to fire a mortar bomb that had six charge increments clipped on and the base plate of the mortar was deep in the mud with the barrel at a shallow angle. With this elevation and charge they were probably hitting Argentine positions around Stanley at the maximum range of 3.6 miles (5,800m).

Suez

In the landings at Suez in 1956 a Royal Artillery naval gunfire forward observer watched with fascination as a Commando rocket launcher crew from 40 Commando fired at the heavy front doors of a large seaside villa. With superb accuracy the crew hit the lock. Following the explosion a small hole was drilled through the lock and the door swung open; inside, the jet of gas from the shaped charge warhead had pulverized the lower half of a huge ornate mirror in the hall. In the smoke the officer could see the undamaged upper half of the mirror sliding slowly down inside the gilded frame.

Suez in 1956 saw a mixture of World War II veterans and National Service Marines in action. One troop commander remembered tough veterans treating house-clearing operations in the town of Port Said as if they were live firing exercises. With rounds cracking overhead and smoke drifting in the streets, they would give the young Marines a quick debrief and bring out lessons learned, before they moved on to clear the next building or block of flats. Here the tactics of fighting in built-up areas (FIBUA) came into their own. The aggressive, fast-moving Marines grenaded buildings before the 'entry men' raced into the smoking wreckage to rake the interior with automatic fire.

Whether holding ground or advancing to contact, the Royal Marines, like the British Army, used patrols to gather intelligence or dominate no man's land. There were three main types: standing patrols, fighting patrols and reconnaissance patrols. Standing patrols could be part of a routine in defence and entail a small force visiting locations that might be used by hostile forces as a forming up point for an attack, an observation post, or might simply be vulnerable to sabotage. What was essential with standing patrols was that their timing should not be predictable.

Fighting patrols could be at troop strength and have various aims. One might be to grab a prisoner for intelligence, or to keep the enemy off balance, to force his patrols back so that they could not gain intelligence or dominate no man's land. The plans for a fighting patrol would have been rehearsed and the standard operational procedures (SOPs) for events like an unexpected enemy contact or white light illumination would have been rehearsed.

The Indonesian Confrontation

The most effective examples of driving in the enemy were the highly classified Claret operations, cross-border raids in the Confrontation against Indonesian military bases. The range and depth of the strong fighting patrols increased as the Indonesians withdrew deeper into Kalimantan. In mid-May 1966 nearly 200 men of L and M companies of 42 Commando crossed the border and marched for two days until they were close to the huts of an Indonesian military base. As dawn approached Claymore mines attached to bamboo poles were pushed silently forwards to fire downwards through the roofs into the huts. At dawn they were detonated and the area lashed with GPMG and Armalite fire. The Marines suffered one casualty, Captain Ian Clark, who died of his wounds.

On 8 December 1964 the men of 40 Commando and the Tawau Assault Group (TAG) received an unusual Claret mission. Set up in 1963, TAG consisted of civilian cabin cruisers, dories and native longboats as well as SBS Klepper canoes manned by men of Nos. 2 and 6 Sections SBS with 40 Commando. They patrolled the Rajang and Kalabakan rivers and the island of Sebatik. The forces were to neutralize an Indonesian observation post (OP) and watchtower on the south-east shore of Sebatik Island that was close to a base on Nanukan Island. What distinguished the mission was that in an earlier century the Dutch and British colonial cartographers had divided Sebatik, an island about 40km by 20km in size, almost exactly in half. The southern portion was now Indonesian and the northern was part of Sabah.

A Fleet Air Arm Sea King helicopter flies over a Bv 202E of 45 Commando as it drives out of Port San Carlos on 21 May 1982. The Sea King HC4 could carry 28 fully equipped men or an underslung load such as a light gun. In training Fleet Air Arm crews work closely with the Royal Marines in challenging and hazardous environments. (RM Museum, 7/20/18 (130))

The SBS had reconnoitred the OP in 1964 and on 8 December that year the motor cruiser *Bob Sawyer* carried a group of Royal Marines under command of an SBS officer, Lieutenant R.A.M. 'Ram' Seeger to a point off the coast. Here they launched three Gemini inflatables. One Gemini containing a support party with a GPMG moved to a position on the border, to deter any follow-up by the Indonesians after the attack. The other two with 15 men aboard paddled the nine kilometres to a small beach. Seeger had divided the 15 men into a GPMG crew under Sergeant Costley, a close-quarter assault group with 9mm Sterling submachine guns under Corporal Tomlin and two scouts with rifles fitted with an improvised night vision device – an electric torch taped to the stocks of their Armalites. As the two inflatables reached the shore the Indonesians opened inaccurate fire. The Marines put in a quick assault with supporting fire from the GPMG on the shore. Ranges were so close that when Seeger gave a warning shout of 'Grenade!' and threw two M26 grenades into the OP, the GPMG crew ducked under water to avoid the blast – but kept the gun above their heads out of the water. Seeger raked the building with SMG fire, and in the light of the campfire the Marines spotted three bodies. With orders not to be drawn into a sustained firefight, the Marines withdrew. Soon afterwards the Indonesians began to mortar the area of the OP, but by then the raiders had opened the throttles on the outboard engines of their Geminis and were out of danger.

This was not the only Claret operation led by Lieutenant Seeger. While serving with 40 Commando he had undertaken two reconnaissance patrols in October 1964. On 12 September 1965, during Operation *Freefall*, D Company of 40 Commando expended over 900 rounds of 7.62mm, 200 rounds of 9mm, 100 rounds of 5.56mm ammunition and eight grenades in a fighting patrol that saw engagement ranges as close as 40m. In the after-action report the attack, which was initiated by Seeger with his Armalite, was described crisply as a 'good operation'.

A GPMG crew of 'F' Company 42 on Mount Kent in June 1982. The men wear the fur pile DPM caps and arctic smocks; the lack of snow meant that conventional camouflage was effective. With a rate of fire of 8,000 to 9,000 rounds a minute, the belt-fed L7A1 GPMG remains a formidable weapon around which a defence or support and attack could be built. (RM Museum, 7/20/18 (133))

Reconnaissance or recce patrols were 'sneaky beaky' operations in which the aim was to see and not be seen. While they are often seen as the speciality of the SBS, all Marines are trained to carry out recce patrols. They could consist of two to six men who might set up a camouflaged OP and use optical and night vision equipment to identify positions and defences. Information might be brought back in note form, dictated into a hand-held recorder or transmitted by radio using codes or encryption. In order to ensure that men obtained rest and thus remained alert, a long-standing OP needed to have enough men to allow two to stand down and rest. This type of work required patience and stamina.

In the Confrontation the tactic most favoured was the close-range ambush. Narrow jungle trails with bends and blind corners or jungle clearings made ideal 'killing grounds', the area in which an enemy force could be trapped and destroyed by overwhelming close-range fire.

There are two types of ambush: deliberate and immediate. The immediate ambush is sprung when an unsuspecting enemy has been spotted approaching and the patrol pulls off the track into the jungle to catch them as they pass; the action is over in minutes. Deliberate ambushes, often based on intelligence, require planning, briefing, orders and rehearsals and may be set up and remain in place for a couple of days or longer. The ambushers must work out a routine in which some withdraw from the ambush site to a secure area to rest and eat; however, food preparation using fires is unwise, since the smells might alert the enemy.

The tactics that were developed in the Confrontation were to catch an Indonesian raiding party as it returned to the safety of Kalimantan, Indonesian Borneo. Since there were identifiable routes back the border ambush groups would be inserted by helicopter near these tracks and sometimes if they were

A ROYAL MARINE IN THE FALKLANDS CAMPAIGN, 1982
Much of the clothing and equipment developed for winter operations on the northern flank of NATO in Norway came into its own in the Falklands in 1982. The hooded DPM arctic smock and trousers with their large buttons, DPM mitts, fingerless gloves, headovers, gaiters (made from breathable material), and boots ski/march were among the kit used by Royal Marines. One added advantage for the Royal Marines involved was that the men had trained for this harsh environment. On the journey south there had been some light-hearted discussion about what type of wax would be suitable for skis; in the end the snow that fell was only light, though wind and rain made conditions miserable. This Royal Marine, pictured outside Government House in Stanley at the end of the campaign, is wearing olive drab 1958-pattern webbing over his jacket, and is armed with an AR-15 Colt Armalite rifle.

lucky they would put in an immediate ambush against the raiders. For a deliberate ambush the men would take up several positions. Covering the killing ground was the assault group, and, in an ideal ambush site, the killer group with automatic weapons like the GPMG or LMG would be positioned at right angles on a bend so that they were shooting straight down the track. At either end of the designated killing ground were 'cut-offs'; these groups were positioned to catch the enemy if they doubled back or ran through the ambush. Beyond the cut-offs lay the warning group, to alert the ambush parties. In depth would be the command group, and protection and warning groups would cover the rear of the ambush. On the far side of the killing ground, where the enemy might run for cover, panji stakes (sharpened bamboo driven into the ground at an angle) were set in the jungle, ready to impale them.

An American weapon that was put to good use in ambushes was the command-detonated, directional US M1A Claymore anti-personnel mine. A compact curved plastic mine with small folding metal legs, it blasted 700 ball bearings across a 60-degree arc at a height of two metres. It was lethal up to 164ft (50m) and could cause casualties up to 328ft (100m). The Claymore could be positioned at ground level, but was light enough that it could even be mounted in trees and even mounted on the end of bamboo poles like a modern pole charge.

In the mid 1960s in Borneo the use by men of 40 and 42 Commando of another American-designed weapon, the Armalite rifle, was innovative. It had been designed by Eugene Stoner as an aircrew survival weapon for the USAAF and was made from alloy and plastic. It fired a 0.223-inch (5.56mm) round with a high muzzle velocity of 3,199f/s (975m/s) (the SLR has a muzzle velocity of 2,799f/s (853m/s)), but most importantly for men loaded with kit and equipment it weighed only 6.4lb (2.9kg).

Aden

In the May 1964 fighting in the Radfan Mountains in the mountainous border of the British-administered Western Aden Protectorate, the men of 45 Commando relearned the tactics of mountain warfare in a harsh climate – one that would be revisited in the 21st century in Afghanistan. Helicopters gave greater tactical mobility, allowing men to picquet the heights and so dominate the area, but they were a limited resource and mobility often depended on the stamina of Royal Marines – loaded with weapons, ammunition, water and rations – who had to climb to the top of sun-blasted features. However, the men discovered that it was possible to move by night, which was quicker and cooler, using the low ground. The golden rule, though, was that troops should have reached the high ground by first light. Men learned the value of requesting, and controlling, artillery and air support.

As always the outstanding tactical lesson of the campaign was the need for the highest standards of junior leadership. It was a campaign fought primarily by troops and companies often operating in isolation in difficult country, where they were obliged to rely on their own resources to deal with unexpected trouble.

The rivalry between the Royal Marines and Parachute Regiment – with 3 Para operating in the Radfan in May – took on an unusual twist on 26 May. The helicopter carrying the CO of 3 Para, Lieutenant-Colonel Antony Farrar-Hockley, was hit by small-arms fire and forced to land near Z Company positions. The word went about that the redoubtable Para colonel had been rescued by the Royal Marines.

The Falklands

In the Falklands in 1982 the Mountain and Arctic Warfare Cadre fought a very effective patrol action at Top Malo House against an Argentine special forces patrol, and then went on to conduct some testing recce patrols on the Argentine-held features that were Commando objectives. In one recce patrol against Two Sisters, the feature that was the objective of 42 Commando, two men, Sergeant Wassell and Lieutenant Haddow, had remained completely still on an exposed area of moorland for a whole day. At a range of only 50 yards (45.7m) they watched the Argentine garrison on Tumbledown and Two Sisters and plotted the command-detonated mines sited on the eastern peak of Two Sisters. After last light both men then returned with a detailed description of the objective.

The skills of amphibious warfare, which many pundits had long seen as redundant, were crucial to the success of Operation *Corporate*, the liberation of the Falkland Islands. The Royal Marines felt a special bond with the islands, since Naval Party 8901 had provided the tiny garrison before the Argentine invasion.

The British landings, which began at 02.30 hours on Friday 21 May, saw 2 Para and 40 Commando in the first wave to reach the shore at the little fishing settlement of San Carlos. 40 Commando would remain to hold the community as the vital logistic base for subsequent operations on the island. Following the casualties amongst the Welsh Guards, men of 40 Commando would join the battalion as reinforcements for the final attack on Stanley.

The tough training standards of the Royal Marines paid off in the Falklands, when following the loss of vital Chinook helicopters aboard the container ship *Atlantic Conveyor*, sunk by Argentine air-launched Exocet anti-ship missiles on 25 May, the men of 45 Commando 'yomped' across the north of the island. This gruelling cross-country march was undertaken at the same time that 3 Paras 'tabbed' the distance to be in position for the assault on the hills to the west of Stanley.

The attacks in the Falklands were unlike anything that many of the men of 42 and 45 Commando had experienced before. This was conventional war with support from fighter ground-attack aircraft, artillery and naval gunfire, and these assets would be combined into fire plans that backed up the ground assaults on the mountain features covering the westerly approaches to Stanley.

The objective of 42 Commando would be a feature known as Mount Harriet, while that of 45 Commando

A Royal Marine of 40 Commando searches an Argentine soldier at Port Howard at the close of Operation *Corporate* in 1982. The Marine wears a headover against the cold, but with hostilities at a close does not have his webbing. He retains a loaded SLR for self-protection. (Al Campbell)

A patrol of 42 Commando in Port Stanley takes time to talk to local youngsters. One man is armed with a 7.62mm light machine gun and the other with an SLR. In the background the police station shows the damage to the roof from a hit by a helicopter-launched anti-tank missile. (Pete Holgate)

was Two Sisters to the north. The features were held by men of the Argentine 4th Regiment, which, though it consisted largely of conscripts, had a backbone of regular NCOs and officers. Crucially, they had good positions and time to prepare defences. Minefields had been laid, but they had neglected to put out barbed wire – a simple obstacle that would have hindered or even stopped some of the attacks. The attack by the two Commandos would go in on the night of 11–12 June 1982.

The attack on Mount Harriet used the powerful element of surprise. While J or 'Juliet' Company would provide fire support from the mountain codenamed 'Tara', L ('Lima') and K ('Kilo') companies would make a long night march of over five miles to be in position to assault Mount Harriet, codenamed 'Zoya', from the east. One position put up a spirited resistance, but Corporal Steve Newland, though wounded in both legs, grenaded the position before charging in with his Sterling SMG. The position finally fell when Corporal Chris Ward fired a 66mm anti-tank weapon at it. The shock and surprise paid off; the Commando lost two junior NCOs killed and 17 men wounded, and took 300 prisoners.

On Two Sisters 42 Commando launched a two-phase attack with X ('X-ray') Company taking a feature to the south and west of Two Sisters. It was a tough fight, with the company sergeant-major controlling fire support from seven LMGs and the 66mm light anti-tank weapons (LAWs).

Soon after midnight on 11 June the main weight of the attack made up of Y and Z companies fell on the northern flank of Two Sisters, which was codenamed 'Summer Days'. For the Argentine garrison the attack by Z Company was signalled by the war cry of 'Zulu! Zulu!' Julian Thompson, then commanding 3 Commando Brigade, describes the action thus in his book *No Picnic*:

> The combination of good control, fitness and the proper use of fire power had enabled them [45 Commando] to take this formidable position with light casualties. Their commanding officer's calm voice on the radio had been like a tonic to all who heard him.

The fight for Two Sisters had lasted 2.5 hours and cost the Commando four killed and 11 wounded. They had taken the highest feature in the mountains west of Stanley and with it 44 prisoners – the remainder had fled, leaving ten dead behind them. Only men who were well trained and led could have fought and won this action against an enemy holding an ideal defensive position.

Prior to the British landings, a potential armoured threat had been identified from French-built Panhard AML90 armoured cars, and so the support company of 40, 42 and 45 Commandos had brought their Milan wire-guided anti-tank weapons (ATGW) to the Falklands. The men had trained hard on simulators on the journey south and worked out how the missile and firing post could be man-packed across country. In the attack on Mount Harriet by 42 Commando the Milan, which had been brought forward at night, was used against enemy machine-gun positions on the south side of the feature. The missile it used had a 3 lb (1.36kg) HE shaped-charge warhead and the two-stage rocket motor took 12.5 seconds to reach 1.2 miles (2,000m).

Conclusion

The widely varying combat experiences between 1950 and 1982 have tested the mettle of the Royal Marines. The Corps has seen action in very diverse theatres and environments, from fighting in the grim Korean winter of 1950/51, in the barren sun-scorched mountains of the Radfan, through the dark and damp jungle of North Borneo, on the hostile streets of Northern Ireland, to the deployment of 3 Commando Brigade in the Falklands in 1982. The tough selection process that recruits had to pass through to win their green berets ensured that the Royal Marine Commandos were fit and adaptable enough to take on these diverse challenges.

In a world in which military operations are now increasingly complex and the distinction between fighters and civilians harder to discern, the qualities of stamina, courage, flexibility and quick judgement are essential for soldiers on the ill-defined front line. In the 26 years between Korea and the Falklands, the Royal Marines have demonstrated they have these strengths, and have now taken them into the new conflicts of the 21st century.

MUSEUMS AND COLLECTIONS

For the researcher or interested visitor, the superb Royal Marines Museum in Southsea, Hampshire is a must. The figures, artefacts, films – and even a live snake in the display about jungle warfare operations – make this a unique military museum.

Royal Marines Museum
Eastney Esplanade
Southsea PO4 9PX
Hampshire
(website: www.royalmarinesmuseum.co.uk),

Additional information about the Royal Marines and latterly the Commandos can be found at the following institutions and museums:

The Imperial War Museum
Lambeth Road
London SE1 6HZ
(website: www.iwm.org.uk)

National Maritime Museum
Greenwich
London SE10 9NF
(website: www.nmm.ac.uk)

National Army Museum
Royal Hospital Road
Chelsea
London SW3 4HT
(website: www.national-army-museum.ac.uk)

Collecting

While Royal Marine uniforms and insignia do come on the market, it is often the ceremonial items that attract the most interest. One source for badges and insignia is Ian Kelly Military Insignia (website: www.kellybadge.co.uk). A brief examination of Ebay will turn up items such as a Royal Marine green beret and bronze cap badge, a pair of 'Royal Marine Commando' shoulder titles, a Falklands vintage reversible green and white waterproof, an officer's 1950s battledress and a new Arctic windproof smock – and many others.

For a collector looking to reconstruct a figure from the period covered by this book it is worth remembering that much of the combat equipment used by Commandos was standard to the British Army. Exceptions were the arctic warfare clothing used by Commandos in Norway on the NATO flank.

More informal items related to the Royal Marines and Commandos, such as coffee mugs, engraved glasses, t-shirts, knit caps and even teddy bears can be bought at the excellent Royal Marine Museum shop.

BIBLIOGRAPHY

Daugherty, Leo *Train Wreckers and Ghost Killers: Allied Marines in the Korean War* (USMC Historical Center, 2003)

Fowler, William *The Royal Marines 1956–84* (Osprey Publishing Ltd, 1984)

—— *Britain's Secret War: The Indonesian Confrontation 1962–66* (Osprey Publishing Ltd, 2006)

Thomas, Lieutenant-Colonel Peter *41 Independent Commando R.M. Korea 1950–1952* (Royal Marines Historical Society, 1990)

Thompson, Julian *The Royal Marines from Sea Soldiers to a Special Force* (Sidgwick and Jackson, 2000)

Ladd, James D. *The Royal Marines 1919–1980: An Authorised History* (Jane's, 1980)

GLOSSARY AND ABBREVIATIONS

The terms in this glossary are a mixture of the accepted military abbreviations and Royal Marine slang. While some of this slang is used in other branches and arms of the British armed forces and has passed into common usage, some remain unique to the Royal Marines

Ace	Good, excellent.
Animal run	An extravagantly indulgent run ashore.
AWT	Arctic-warfare trained.
Banjo	To kill or destroy.
Bimble	To wander around with no obvious objective, e.g. 'bimbling along'.
Bite	To be drawn into an argument or be fooled.
Bootneck/bootie	A Royal Marine, from the leather uniform stock worn around the collar in the 18th/19th centuries. The USMC equivalent is 'leatherneck'.
Bombed out	Drunk or crazy.
Boss	Informal but respectful mode of address for an officer.
Brill/brills	Brilliant, excellent.
Brammer	Outstandingly good.
Bronzy	A sun tan.
Bug out	To escape, make a military withdrawal.
Buzz	A rumour, or the general description of a situation.
CGRM	Commandant General Royal Marines.
Chuck one up	To salute.
Cloggie	A Royal Netherlands Marine.
Common dog	Common sense.
Corps, the	The Royal Marines.
Crab/Crab Air	RAF personnel or the RAF as a whole, hence the phrase: 'Time to spare, fly Crab Air'.
Crack	To achieve.
Crappers	Very drunk.
Cream in	Collide or crash.
Crimbo	Christmas.
CTC	Commando Training Centre.
Dig out (blind)	To make a supreme effort.
Dischuffed	Offended, opposite of chuffed.
Drip	Complain.
Eighty-eight	Carl Gustav 88mm recoilless anti-tank weapon.
Essence	Beautiful, normally used for a 'pash' or 'party'.
Flakers	Exhausted.
Foo foo	Issue 'foot and body powder'.

Gen	The truth.
Glimp	To peer or peep.
Glophead	A drunkard, from 'glop' – to slurp or drink noisily.
Goffer	A wave, a cold drink or a punch.
Gonk	To sleep.
GPMG	7.62mm General Purpose Machine Gun, often known as the 'Jimpy'.
Grolly	Something unpleasant.
Gronk	An unattractive woman.
Hand	Reliable efficient man, a term derived from the Royal Navy.
Ish, the	To have all available equipment, to be the best.
Kag, kaggage	Unwanted useless equipment or stores.
Knacker	An unfit person. A 'fat knacker' is a fat, unfit person.
KUA	Kit upkeep allowance.
LCA	Landing craft assault.
LCU	Landing craft utility.
LCVP	Landing craft, vehicle/personnel.
LAW	Light anti-tank weapon.
Limers	A soft drink with a high vitamin C content.
LMG	Light machine gun.
Loopy/looney juice	Alcohol.
LPD	Landing platform dock.
LST	Landing ship tank.
LZ	Landing zone.
M&AW	Mountain and arctic warfare.
Maskers	Black or green masking tape, also known as Pusser's tape.
Mankey	Filthy.
Minging	Drunk or filthy.
ML	Mountain leader.
Mod Plod	Ministry of Defence policeman.
Nause	Inconvenience.
NBC	Nuclear, biological and chemical.
Nod/Noddy	RM recruit.
No man's land	The area that separates the front lines of two opposing forces.
Nutty	Sweets, commonly chocolate bars.
Nutty fiend	A man with a sweet tooth.
OP	Observation post.
Party	Female.
Pash	A female with whom a close friendship has been formed.
Pongo/Percy Pongo	A soldier or any non Royal Marine. From 'Everywhere the Army goes, the pong goes too' – based on the idea that men aboard a troopship rarely washed.
Picturize	To give an explanation.
Ping	To recognize or identify, from the anti-submarine Asdic 'ping'.

Piso	Careful with money.
Plums rating	A man who is always unsuccessful with women.
Prof	To do well, or acquire.
Pussers	Anything (equipment, stores, regulations) to do with the Royal Marines or Royal Navy, or anything done unimaginatively or according to regulations.
Puzzle palace	Headquarters building.
Rice	Effort, hence the exhortation 'Give it rice'.
RM	Royal Marine.
Rock-all	Nothing.
Royal	Royal Marine.
Run ashore	Time off away from ship, barracks, base or office.
Sad-on	Unhappy, hence 'Don't get a sad-on…'
SC	Swimmer canoeist.
Scran	Food. Contrast with the Army 'scoff'.
SF	Sustained fire.
Shave off	Expression of annoyance, or to speak out of turn.
Shoot-through	Someone who fails a task.
Sixty-six	66mm M72 light anti-tank weapon.
Skeg	To observe or make a reconnaissance.
Skin	An inexperienced or immature young man.
SLR	L1A1 7.62mm self-loading rifle
Sneaky beaky	Intelligence staff and their operations.
Snurgle	To approach cautiously or crawl, thus 'I snurgled up.'
SOP	Standard operating procedure.
Sprog	Child, or inexperienced man.
Stacks	Opposite of plums, especially in relation to women.
Thin-out	Depart, can be used as an instruction 'OK, thin out…'
Threaders	Fed up.
Trap	To successfully attract a female.
Trough	To eat.
Twitter	To talk aimlessly.
UN	United Nations.
Up homers	Invited into a home.
Up the line	Travel away from base (this phrase dates back to World War I).
Wet	A drink, normally hot. Army term is 'brew'.
Wrap	To stop or give up.
Yaffle	To eat hurriedly.
Yeti	A spectacular fall while skiing.
Yomp	Cross-country march carrying a heavy load.
Yo-Yo	A young officer under training.
Zap	Shoot or be shot.
Zeds	Sleep, hence ' I'm going to shoot some zeds' – from the cartoon 'Zzzz…' to indicate sleeping.

INDEX